Critiquing the DSM 5

2015

Critiquing the DSM 5

Compiled by Nora L. Ishibashi

2015

First Printing: 2015

ISBN 978-1-329-19880-7

Nora L. Ishibashi, Ph.D., P.C.
405 North Wabash Avenue, #1303
Chicago, IL 60611

www.noraishibashi com

Ordering information:
This book may be ordered from http://www.lulu.com

Table of Contents

Introduction

The assessment and diagnosis of psychological distress requires a sophisticated understanding of human behavior, dysfunction, and treatment. The *Diagnostic and Statistical Manual of Mental Disorders, Fifth Edition*, is the catalog of diagnostic categories most commonly used by clinical professionals treating these disorders. These papers represent the thoughts and responses of students in the master's degree program at Loyola University of Chicago to the current practice of assigning diagnoses to psychological symptoms. In our course covering assessment and diagnosis, we consider current practices in the field of social work and apply theoretical views in order to critique and learn from those practices. An important goal of our profession is continuing to question and improve our knowledge base and our methods of being helpful to the people for whom we care.

Understanding the complexities of another person's life requires perceptive attention and sophisticated analysis. Often we have only the outlines of the story from our conversations with our clients. We attempt to discover what may have precipitated the distress, what it might have in common with other kinds of distress we have seen and what might be different. We design responses which we hope will be constructive and effective. Before we can do that, we must first identify the difficulty.

Social work is a profession with a history of more than one hundred years and a complex array of theories and methods. We are committed to finding ways to be helpful to people who may be struggling with circumstances or destructive events knowing that at one time or another each of us will take turns being helpers and being helped. We follow a long line of professionals who have worked hard to provide care, to understand and express the ways people struggle and to provide the kinds of help that are useful.

You will find in these papers not only the thoughtful ideas of a new generation of social workers, but also a sense of the very altruistic and intelligent people who are committing their lives to caring for others

Debunking the Diagnosis of Major Depressive Disorder

Brian Ahern

Introduction

When a traumatic event occurs in a child's life, such as the loss of a loved one, the bereaved child may go through long and intensive mourning. Children often exceed the "normal" time allowed to grieve properly, which is a period that is needlessly determined by society. Research has shown that it is these children who are deemed outliers of typical development, and thus unnecessarily given the label of Major Depressive Disorder (Burke, 2003). This chapter will try to denounce the diagnosis of Major Depressive Disorder by showing that the "typical grieving process" is a myth because grief is a process that cannot nor should not be measured. This is especially true for children in poverty who do not have the resources to work through mourning a loved one, as grieving may be a process for the entirety of their lives. In this chapter, the manifestations of grief will be looked at through the lens of Ecological Systems Theory, as all one has to do is look at the systems that shape a child's development in order to understand his or her behavior, rather than confining that child in a box when labeling him with Major Depressive Disorder.

A Definition of Grief

For the sake of this chapter, grief will be defined as deep sorrow, caused by a loved one's death. In addition to this definition, due to grief not being measurable, the chapter also views the grief process as not having a timeline. Rather, due to the intense nature of a traumatic loss, and depending on the supports in a child's life affecting how a child expresses his emotions, this chapter suggests that it would be very natural for most individuals to be in a constant state of grief for their whole life. The author of this chapter has direct experience working with youth, specifically leading grief and loss groups for fifth and sixth graders in the Austin neighborhood of Chicago, IL. Like any other characteristic, grief looks different in all children due to all children being unique individuals. Various manifestations of grief in children are resiliency, generalized fears about death, specific phobias, depressive symptoms, behavioral issues, and hyperarousal symptoms (e.g. sleep disturbance and angry outbursts) (Killen et al., 2007). Grieving can also look different depending on the type of loss one has experienced. Some of these factors include the severity of the loss, how close the griever was to the person whom was lost, and the nature of the loss (e.g. sudden, unpredicted, enduring, or recurring) (Dyregrov et al., 2013).

Different Manifestations of Grief in Children

What makes grief complex is that it is manifested in many different ways, depending on the systems in the child's life that have given the child the resiliency or risk factors to work through the grief (Dyregrov, 2013). In this author's work with children in the grieving process (e.g. middle school boys in urban poverty), particularly in leading group psychotherapy sessions for children who have lost a loved one, there have been extensive experiences seeing these different manifestations play out in the school setting. These experiences, and subsequent research on grief, has led to the definition of four emotional states that children tend to fall into when mourning the loss of a loved one. It is important to note that the behavioral states of children listed below are used in describing behaviors that in the past have been unnecessarily categorized as Major Depressive Disorder, and that these states are certainly not exhaustive, as there are children who will exhibit grief in other ways. By these states being subsequently defined from experiences working with males, it should be noted that these states have been identified for males in particular, although females may fall into these categories as well[1]. It is also imperative to understand that these states are not mutually exclusive, and a therapist could find the child in two, three or all four states at once, depending on how traumatic the loss was for the child. The four proposed states of grief are as follows: Proactive, Inactive, Active, and Reactive grieving states. Diagram 1 provides a visual representation as to how these states relate to each other, as they all have a commonality in traumatic loss.

---

[1] The author of this chapter has no experiences working with middle school females, thus for the sake of this chapter, has chosen to focus on grief in middle school males.

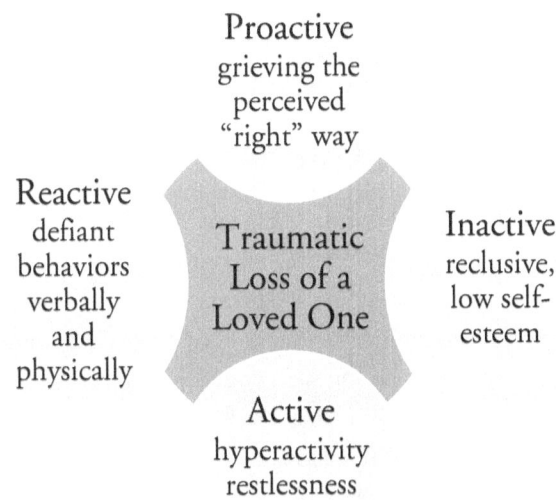

Proactive
grieving the
perceived
"right" way

Reactive
defiant
behaviors
verbally
and
physically

Traumatic
Loss of a
Loved One

Inactive
reclusive,
low self-
esteem

Active
hyperactivity
restlessness

States of Grieving in Children

Proactive Grieving

Proactive grieving involves the child talking through the traumatic loss and expressing emotions in a behaviorally positive way. This is the child who is generally perceived by society as grieving in the "right" way. This child typically has support systems in place (e.g. family members, peers, school, outside resources) to help cope and guide him to appropriately heal from the deep wound that has occurred. Not without sadness, this child is taking the right steps to using his traumatic loss to influence his life, as he assimilates back into a daily routine. This child may not need any outside grieving services, but this does not exclude him from participating in these entities.

Inactive Grieving

The child in the Inactive state of grieving shows signs of extreme sadness due to the loss of a loved one. This child may talk about his loved one, but divulging sensitive information may elicit gloomy and reclusive behaviors. The child may not actively seek out opportunities to talk about his loss, but if reached out to, he may be more than willing to talk than other grief states (e.g.

Active and Reactive). The emotions of a child in Inactive grieving do not fluctuate as much as the other states, as sadness is predominately present. Notably, this child may remove himself from his peers, due to low self-esteem and an inability to make friends easily.

Active Grieving

The child who is in a state of Active grieving is physiologically hyperactive, and is on the opposite end of the spectrum compared to the child in Inactive grieving. He is visibly nervous to be in a situation, such as individual or group psychotherapy, in which the intent is to elicit emotions about the loss of a loved one. This child can be seen constantly fidgeting around. The fidgeting can result in a distraction to the student from being present, but it more notably results in fidgeting that disrupts the rest of the group. This fidgeting can also be seen verbally, as the child is so uncomfortable in sharing about his loss, he may bring up a myriad of different topics in order to avoid talking about his feelings. This child also may provide excuses as to why he does not want to talk about his loss, which are usually physiologically-based (e.g. "my head hurts too much").

Reactive Grieving

The child in the Reactive state of grieving exudes defiant behaviors and attitudes towards anyone who tries to elicit any emotions that are related to addressing the loss he experienced. These defiant behaviors are manifested by sayings such as, "This is stupid" or "I am not going to do this," often times denying that there is anything that he needs to talk about. When the child speaks of the loved one who passed away, it is often malicious and involving the child blaming that person for things that went wrong when they were alive. Additionally, this child does not have to say anything in order to show his defiance, as ignoring the therapist's directions and his peer's reflections in the group can be a common occurrence. A child in the reactive state may also take his defiance out physically, which can result in walking out of sessions or making fun of group members to a point where there is a physical altercation.

Diagnosis of Major Depressive Disorder
Given to Grieving Children

Major Depressive Disorder (MDD) is the diagnosis that has been chosen due to it commonly being given to children who are bereaved and still in the grieving process, which as discussed earlier has no timetable. According to the Diagnostic and Statistical Manual of Mental Disorders Fifth Edition (DSM 5), a diagnosis of MDD includes multiple symptoms that have to be consistently present and subsequently impairing in the individual's life (APA, 2013). These symptoms include depressed moods, diminished interest or pleasure in all or almost all activities, significant weight loss, insomnia, psychomotor agitation or retardation, fatigue or loss of energy, feelings of worthlessness or inappropriate guilt, diminished ability to think or concentrate, and recurrent thoughts of death or suicidal ideation (APA, 2013). It is important to note that five or more of these symptoms must be met nearly every day in the two-week span, for the majority of the day (APA, 2013). This diagnosis also specifically targets children when stating that irritability can be a replacement for constant sadness, as a symptom (APA, 2013).

Consequences of Major Depressive Disorder Symptoms

According to Hazell (2002), children who show signs of MDD face a myriad of symptoms including irritability, lack of interest in activities they previously enjoyed, difficulties with concentration, lack of energy, sleep issues, and overt pessimism. Black (1987) also mentions children showing depressive symptoms such as anxiety, eating disturbances (e.g. overeating, undereating), school refusal (e.g. not wanting to go to school, not wanting to do work while at school), and inappropriate obsessions (e.g. hoarding). The DSM 5 adds that there are varying levels of impairment, with some symptoms so mild that other individuals may not even be aware of them (APA, 2013). Particularly in the school setting, symptoms of MDD could potentially be seen in the student avoiding conversations with people, subsequently ensuring that they don't have a lot of friends (APA, 2013). Symptoms of MDD can also manifest in performance problems in school, altercations with peers, and denial of emotions mixed with aggression towards mental health professionals, as these children want to be independent, and they think they can go through the grieving process alone (APA, 2013).

Issues with Diagnosing Major Depressive Disorder in Children

There has been recent literature describing the difficulties in diagnosing children with MDD, for various reasons. McDermott (1980) tested the consistency of school mental health services and found that no matter what level of experience one had (e.g. novice, intern, licensed psychologist), that consistency in diagnosing sample case studies was considerably lacking, and even as these professionals acquired more training, further disagreement on diagnoses increased. This points out the subjectivity in which mental health professionals give labels to their patients, particularly children. According to Murphy (2004), the issue is due to "the symptoms of depression in children are similar to the symptoms of anxiety, attention deficit hyperactivity disorder, and bipolar disorder" (p19), in which "difficulty concentrating, restlessness, poor school performance, and impaired peer relationships can occur in any of these disorders" (p28). Van Schalkwyk (2014) states the issue of diagnosing a child with MDD is that adult psychiatrists do not understand the clients with whom they work. In particular when diagnosing children, psychiatrists are seemingly ill-informed when it comes to recognizing the different manifestations of grief throughout the developmental process. It is especially important to remember that in many settings, official diagnosing occurs as a way to justify reimbursement and hospitalizations (Van Schalkwyk et al., p917, 2014), which can cause hastily created diagnoses, especially when the psychiatrist is under pressure to provide an "answer" in a short amount of time.

An Ecological Systems Theoretical Perspective on Major Depressive Disorder

Looking at the wrongful diagnosis of Major Depressive Disorder through the lens of Ecological Systems Theory is a unique way to look at how the systems in a child's life directly affect how he grieves after a traumatic loss. This perspective shows that the traumatic loss in a child's life is not the only piece that affects a child's grieving process, rather how the child grieves is dependent on the risk and resiliency factors in the child's life. Based on Ecological Systems Theory, the child is supported (and negatively affected) by the systems in place (on varying levels) throughout his entire life. Family, peers, schools, jobs, the economy, and federal policies are but a few examples of various types of systems that directly determine how a person develops throughout life and subsequently determine how they grieve. Ecological Systems Theory is based on the, "idea that an effective system is based on individual needs, rewards, expectations, and attributes of the people living in the system" (Simmons, 2014). Ecological Systems Theory will thus be used to venture into the different levels of a child's life, and to debunk the diagnosis of Major Depressive Disorder given to children, as well.

Microsystem Look at Major Depressive Disorder

The microsystem in the child's life includes the people that are directly involved, and potentially most impactful, in the child's development. This most notably includes the child's family, in particular those adults in the child's life who are raising the child. For example, when a traumatic loss occurs in a child's life, how equipped is the family to help battle the lasting effects of the loss on the child's development? How capable is the family to help the child effectively and supportively cope with the loss? There are many reasons that a child's microsystem would not be able to help the child cope effectively, thus forcing the child's grief to stay unattended. If the child has had multiple traumatic events or losses happen in his life, this can exacerbate the effects of not properly dealing with grief. If the child has to take on more responsibility due to the death of a loved one, the new responsibility of taking care of younger siblings or elderly adults could expedite the grieving process inappropriately. If the role models in the child's life themselves do not give enough attention to the grief process, then the child will not learn otherwise. Other familial factors that directly impact a child's development, thus how a child grieves, are, "adequate routine, including sleep and opportunities for physical exploration/play; and quality of parent–child interaction" (Kirkorian et al., 2009). Cohen (2004) puts it incredibly well in stating:

> "Following a familial death, children may experience secondary adversities such as the loss of home, health insurance, or family income. If the family has to relocate, children may also have to leave their school, peers, place of worship, and other social supports. In these situations, children and parents have to adjust not only to the loss of the loved one, but also to these additional losses" (p819).

Additionally, the family's socioeconomic status can play a part in being a risk or resiliency factor for the child, especially considering, "research has shown that health behaviors change as a function of education level" (Gorski, p5, 2010).

Mesosystem Look at Major Depressive Disorder

In the mesosystem, one has to look at the direct resources available to the child, outside of the family. Does the child have healthy friendships with peers? Does the child have access to therapy? Does the child receive supportive services at his school? Not only is the availability of these resources crucial to the grieving process of a child, but the consistency with which these resources are given is equally important (e.g., the time the child actually spends with his parents compared to inconsistent caretakers, whether insurance will cover the cost of an outside therapist, if the child's school has a social worker, if the school social worker even has enough room on his or her caseload to take on the child). If the child does not have the opportunities to grieve properly through trained professionals or consistent support networks that the child can relate to, then he will continue to improperly grieve which will continue to affect him in a negative way. Peers in the school setting also play a big part in the child's grief development, as peers teasing a child about his deceased family member can be a constant reminder of the loss (Kirwin, 2005), and depending on how traumatic the loss was, and the balance of risk and resiliency factors that the child has internalized, this lack of connection with his peers could exponentially prolong the child's grieving process.

Exosystem Look at Major Depressive Disorder

The Exosystem in the child's life can be described as the social settings which indirectly affect the child's development (e.g. a parent's job requiring them to work more which results in the parent being around the house less). This system is particularly prevalent regarding the socioeconomic class that children are in. For example, if a single mother has to pick up new or extra shifts at her job, and thus has to rely on other people to provide child care, this can drastically effect the development of a child who now has very different rules to live by. Especially among the students that this chapter's author has worked with, some of the neighborhoods in which the students live also have higher amounts of crime and gang involvement. This could result in higher exposure to traumatic violence and possible death of support figures, which could then decrease the amount of supports in the grieving child's life. The schools in neighborhoods of lower socioeconomic status also often have less social-emotional resources (due to lack of funding) such as the consistent presence of a social worker, to guide the child in properly grieving the traumatic losses that he has experienced.

Macrosystem Look at Major Depressive Disorder

The macrosystem is the level in which all systems are embedded. Encompassing all of the other systems, the macrosystem looks at how the culture of the population affects the other systems, which in turn affect the individual. One piece of the macrosystem that seriously hinders the grieving in males is the societal pressure to conform to traditional masculinity. In particular, "Research in the field of men's health and help-seeking has repeatedly indicated that adherence to traditional masculine norms is a negative factor in the physical and mental health of men" (Gorski, p1, 2010), in which the idea is held that as a male you have to hide your emotions, especially feelings of grief and anger (McCreight, 2004). This mentality of holding in your anger, and not expressing your emotions, continues to perpetuate the effect that the traumatic loss has on the child throughout their lifetime. This is the case especially in individual and group psychotherapy. When the male child constantly hears from society, his community, his peers, and his family about how he has to be a "tough man" or the "man of the household," there is no recognition or awareness that the emotions one feels after a traumatic loss are natural nor that these emotions should be processed through seeing a mental health professional. Cheung (2005) touches on this stigmatization in teens, where she found that adolescents are less likely to contact and reach out to mental health services, which results in "perceived negative usefulness of social services and perceived societal disdain for seeking help from the services, affecting students' likelihood of help-seeking in the presence of problems" (Cheung, p63, 2005).

Through television and the internet as well, especially as the use of technology in every aspect of a child's life is rapidly growing (Farber, 2012), the subsequent heightened exposure to the media and America's "pull yourself up by the bootstraps" mentality, is detrimentally affecting how young males feel the need to hold themselves. The message of traditional masculinity is wrongly telling our young men, especially those in lower socioeconomic statuses (e.g. having a greater propensity for lacking positive adult male role models), that if you are strong and don't let things get to you, then you will be successful. This mentality that is forced upon young American men diminishes the necessary process of emotion-sharing, and creates stereotypes and stigmas that they then hold against anyone who does decide to get that extra help with their grieving or emotion building (Gorski, 2010).

Ecological Systems Theoretical Implications on Grief Evaluation

Using the Ecological Systems Theory lens, we now have a construct with which to understand the grief that children feel, no matter how that grief is manifested. Ecological Systems Theory offers a holistic approach to supporting the child in recognizing that individuals, especially children, are incredibly susceptible to the messages surrounding death and loss around them. According to Cheung, promoting the image of receiving mental health services and "encouraging adolescents with problems to seek help from the services" (2004, p. 66), is the best approach to relieving anxious feelings about seeking grief support. Psychoeducation inclusively looks at the factors affecting the child, and places importance on spreading knowledge and awareness about the grief that the child is struggling with. Educating the support systems that the child needs, can only help foster the relationships the child has with the people in his life, thus allowing the child to truly dive into the grieving process. Burke (2003) speaks of conditional goal setting being a very important part of the grieving process for those with depressive symptoms, and lays out various treatment options that can be applied to a child who is grieving, such as "parent therapy, floor time, family therapy, and parent-child interaction training" (p264). In the literature on treating grief resulting from a traumatic loss, there is an increasing number of therapists who are reporting just how well family therapy works in repairing the parent-child relationship, thus allowing the child and the parents alike to make true meaning of the traumatic loss (Mason et al., 2009, Kissane et al., 2006, Groot et al., 2010, Carr et al., 2014).

Conclusion

The current structure we have in place of labeling a child with Major Depressive Disorder when they are in the grieving process is as foolhardy as it is unintelligent. Why is it that we are automatically supposed to not let the loss of a loved one, and a traumatic one at that, affect the rest of our lives in some way? Are we not all in a constant state of grieving for those who are no longer with us? Society's "normal" grief period and the diagnoses that come with it when one does not fit the traditional grieving mold, especially in children, is a myth that needs to be dispelled. Only when understanding the systems in a child's life, can you then see the child as a unique individual who has unique grief. Rather than just diagnosing the symptoms, we need to look at all systems that are affecting the child, as too often these children are unjustly diagnosed with MDD, given a label that they have no business receiving, then given a generalized treatment plan that focuses solely on the individual's symptoms and not on the individual himself. These important questions asked above though, partnered with Ecological Systems Theory, are crucial in confronting this issue of diagnosing MDD in children and imperative in understanding that grief is a part of every individual, it may just be a lifelong journey to understand why.

References

American Psychiatric Association, & American Psychiatric Association. (2013). *Diagnostic and statistical manual of mental disorders: DSM 5*. Washington, DC: American Psychiatric Association.

Aral, N. (2006). A comparison of depression in children with and without mothers. *Psychological Reports, 99*(2), 619-29. doi:10.2466/PR0.99.6.619-629

Burke, M. G. (2003). Depression in preschool children. *Journal of the American Academy of Child and Adolescent Psychiatry, 42*(3), 263-264. Retrieved from http://www.sciencedirect.com.flagship.luc.edu/science/article/pii/S0890856709605542?np=y

Carr, A. (2009). The effectiveness of family therapy and systemic interventions for child-focused problems. *Family Therapy, 31*(1), 3-45. Retrieved from http://web.ebscohost.com.flagship.luc.edu/ehost/detail/detail?sid=ad0325bb-409c-4bad-9767a0b6b5127e59%40sessionmgr4005&vid=0&hid=4204&bdata=JnNpdGU9Whvc3QtbGl2ZQ%3d%3d#db=a9h&AN=35830755

Cheung, C., & Liu, S. (2005). Factors underlying junior high school students? seeking help from social services. *Childhood-a Global Journal of Child Research, 12*(1), 55-69. doi:10.1177/0907568205049892

D'Antonio, J. (2011). Grief and loss of a caregiver in children: A developmental perspective. *Journal of Psychosocial Nursing and Mental Health*, *49*(10), 17-20. Retrieved from http://www.ncbi.nlm.nih.gov/pubmed/21846078

De Groot, M., Neeleman, J., Van der Meer, K., & Burger, H. (2010). The effectiveness of family-based cognitive-behavior grief therapy to prevent complicated grief in relatives of suicide victims: the mediating role of suicide ideation. *Suicide & life-threatening behavior*, *40*(5), 425-37. Retrieved from http://web.ebscohost.com.flagship.luc.edu/ehost/detail/detail?sid=11c9d759-7032-4244-b25a1324492db394%40sessionmgr4003&vid=0&hid=4204&bdata=JnNpdGU9ZWhvc3QtbGl2ZQ%3d%3d#db=i3h&AN=55272912

Dyregrov, A., & Dyregrov, K. (2013). Complicated grief in children - the perspectives of experienced professionals. *Omega*, *67*(3), 291-303. Retrieved from http://web.ebscohost.com.flagship.luc.edu/ehost/detail/detail?sid=68fbd7a8-fe15-4bab-b70ac946e51a4c3b%40sessionmgr4002&vid=0&hid=4204&bdata=JnNpdGU9ZWhvc3QtbGl2ZQ%3d%3d#db=tfh&AN=89676566

Farber, B. A., Shafron, G., Hamadani, J., Wald, E., & Nitzburg, G. (2012). Children, technology, problems, and preferences. *Clinical Psychology*, *68*(11), 1225-1229. Retrieved from http://web.ebscohost.com.flagship.luc.edu/ehost/detail/detail?sid=2e596a2e-2c5a-4ccb-a323fdc189f8f1e3%40sessionmgr4001&vid=0&hid=4204&bdata=JnNpdGU9ZWhvc3QtbGl2ZQ%3d%3d#db=tfh&AN=82370640

Gorski, E. (2010). Stoic, stubborn, or sensitive: How masculinity affects men's help-seeking and help-referring behaviors. *Undergraduate Research*, *13*(1), 1-6.

Hatfield, A. J., & Hipel, K. W. (2002). Risk and systems theory. *Risk Analysis*, *22*(6), 1043-1057. doi:10.1111/1539-6924.00272

Hazell, P. (2002). Depression in children : May go unnoticed and untreated. *British Medical Journal*, *325*(7358), 229-230. doi:10.1136/bmj.325.7358.229

Killen, K., Cohen, J. A., Mannarino, A. P., & Deblinger, E. (2007). Treating trauma and traumatic grief in children and adolescents. *Child Abuse & Neglect*, *31*(5), 256. Retrieved from http://www.ncbi.nlm.nih.gov/pmc/articles/PMC2809450/

Kirkorian, H. L., Pempek, T. A., Murphy, L. A., Schmidt, M. E., & Anderson, D. R. (2009). The impact of background television on parent-child interaction. *Child Development*, *80*(5), 1350-9. doi:10.1111/j.1467-8624.2009.01337.x

Kirwin, K. M., & Hamrin, V. (2005). Decreasing the risk of complicated bereavement and future psychiatric disorders in children. *Journal of Child and Adolescent Psychiatric Nursing, 18*(2), 62-78. doi:10.1111/j.1744-6171.2005.00002.x

Kissane, D. W., McKenzie, M., Bloch, S., Moskowitz, C., McKenzie, D. P., & O'Neill, I. (2006). Family focused grief therapy: A randomized, controlled trial in palliative care and bereavement. *American Journal of Psychiatry, 163*(7), 1208-18. doi:10.1176/appi.ajp.163.7.1208

Little, S. G., & Akin-Little, A. (2013). Trauma in children: A call to action in school psychology. *Journal of Applied School Psychology, 29*(4), 375-388. Retrieved from http://www.tandfonline.com.flagship.luc.edu/doi/abs/10.1080/15377903.2012.695769#.VSxhOSvF-So

Mason, J., & Haselau, C. (2000). Grief therapy: An experiential workshop. *Contemporary Family Therapy, 22*(3), 279-288. doi:10.1023/A:1007808523540

McCreight, B. S. (2004). A grief ignored: Narratives of pregnancy loss from a male perspective. *Sociology of Health & Illness, 26*(3), 326-350. doi:10.1111/j.1467-9566.2004.00393.x

McDermott, P. A. (1980). Congruence and typology of diagnoses in school psychology: An empirical study. *Psychology in The Schools, 17*(1), 12-24. doi:10.1002/1520-6807(198001)17:13.0.CO;2-B

Murphy, K. (2004). Recognizing depression in children. *The Nurse Practitioner, 29*(9), 18-29. doi:10.1097/00006205-200409000-00003

Nathan, P., Durkin, K., Morling, J., Dzahari, M. A., Carson, J., & Durkin, E. (2004). Understanding the relationships between wellbeing, goal-setting and depression in children. *Australian and New Zealand Journal of Psychiatry, 38*(3), 155-161. doi:10.1080/j.1440-1614.2004.01317.x

Neal, J. W., & Neal, Z. P. (2013). Nested or network? Future directions for ecological systems theory. *Social Development, 22*(4), 722-737. Retrieved from http://web.ebscohost.com.flagship.luc.edu/ehost/detail/detail?sid=e3707c6d-d99c-4dbd-b183e8398288ddcd%40sessionmgr4004&vid=0&hid=4204&bdata=JnNpdGU9Z Whvc3QtbGl2ZQ%3d%3d#db=a9h&AN=90633545

Simmons Staff. (2014, May 6). *Theories Used in Social Work Practice & Practice Models - SocialWork@Simmons*. Retrieved from http://socialwork.simmons.edu/theories-used-social-work-practice/

Van Schalkwyk, G., Peluso, F., Qayyum, Z., McPartland, J., & Volkmar, F. (2015). Varieties of misdiagnosis in ASD: An illustrative case series. *Journal of Autism and*

Developmental Disorders, *45*(4), 911-918. Retrieved from
http://link.springer.com.flagship.luc.edu/article/10.1007%2Fs10803-014-2239-y

Emotion or Pathology: A Feminist Take on Borderline Personality Disorder

Niki Aquino

Introduction

Diagnosing a person, regardless of the disorder, is to give that person a label based on the observations of a third party, usually a clinician or physician. Although clinicians take into account the client's personal experiences and perspective, the diagnosis is ultimately made by the clinician. This paper intends to focus on the diagnosis of a specific personality disorder, Borderline Personality Disorder, the label it assigns those diagnosed with it, and how it may be viewed through the lens of feminist theory. There are many ways in which a clinician's opinion could and does affect a given diagnosis. In regards to Borderline Personality Disorder, it may be possible to reduce the influence of gender bias on this diagnosis through a careful critique of the criteria and the impact of societal undertones in naming women's emotions and behaviors as pathological.

Defining Borderline Personality Disorder

Borderline Personality Disorder (BPD) first appeared in the 3rd Edition of the Diagnostic and Statistical Manual of Mental Disorders (DSM) and is now "one of the most intensively researched personality disorders – in terms of diagnosis, developmental psychology, pathophysiology, and treatment" (Kernberg & Michels, 2009, p. 505). Key to the understanding of BPD, and all personality disorders, is that "the concept of a personality disorder assumes that there exists a characteristic and pathological personality substrate that will shape the form and content of the psychopathology manifested, and will serve as a source of psychological dysfunction even in the absence of external stressors" (Brown & Ballou, 1992, p. 210). Individuals with personality disorders have inflexible and maladaptive traits that severely impair their lives interpersonally and occupationally.

According to the 5th Edition of the DSM, BPD is characterized by a "pervasive pattern of instability of interpersonal relationships, self-image, and affects, and marked by impulsivity, beginning in early adulthood and present in a variety of contexts" (American Psychiatric Association {APA}, 2013, p. 663). This definition is followed by nine criteria further characterizing the chaos and instability experienced in BPD, although only five are needed to warrant a BPD diagnosis. Many of the criteria are interrelated and share the common theme of instability.

Criteria 1 and 2 encompass the volatile interpersonal relationships of persons with BPD. Persons with BPD constantly fear abandonment and "make frantic efforts to avoid real or imagined abandonment" (Criterion 1) (APA, 2013, p. 663). The feelings of possible separation or rejection are sensitive to environmental circumstances, whether planned and predetermined or unexpected. The individual's panic may be triggered by "realistic time-limited separation or when there are unavoidable changes in plans" (APA, 2013, p. 663). For example, an individual with BPD may experience panic or fury when a lover or close friend is running late or must cancel a meeting last minute due to an unexpected circumstance. They may believe this "abandonment," whether perceived or real, is an indication that they are "bad" and deserve to be left. They experience contradictory feelings of "an intolerance of being alone and a need to have other people with them" (APA, 2013, p. 663). These feelings greatly affect the relationships persons with BPD have with others.

Criterion 2 is "a pattern of unstable and intense interpersonal relationships characterized by alternating between extremes of idealization and devaluation" (APA, 2013, p. 663). Persons with BPD are prone to dramatically shift their view of others. They may cling to an individual, demand to spend extensive amounts of time together, and share intimate details. Then, they may switch and intensively devalue the individual and claim "they do not care enough, give enough, or is not 'there' enough" (APA, 2013, p. 664). Relationships with persons with BPD are contingent upon the promise that the other person will "be there to meet their needs and demands" (APA, 2013, p. 664). The tumultuous nature of interpersonal relationships in those with BPD is matched by internal chaos and conflict

Criterion 3 is an "identity disturbance characterized by markedly and persistently unstable self-image or sense of self" (APA, 2013, p. 664). The individual's unstable sense of self may result in changing goals, occupational interests, political views, and values and morals. Individuals with the disorder usually have self-images characterized as all good or all bad, but sometimes they have "feelings that they do not exist at all" and experience "chronic feelings of emptiness" (Criterion 7) (APA, 2013, p. 664). These feelings often "occur in situations in which the individual feels a lack of a meaningful relationship, nurturing, and support" (APA, 2013, p. 664). In addition, BPD impacts behavior, affect, and mood.

In terms of behavior, Criterion 4 requires "impulsivity in at least two areas that are potentially self-damaging" (APA, 2013, p. 663). Common behaviors include gambling, irresponsibility with finances, substance abuse, unsafe sex, and reckless driving. Another behavior key to BPD is "recurrent suicidal behavior, gestures, or threats, or self-mutilating behavior" (Criterion 5) (APA, 2013, p. 664). This behavior is sometimes displayed in the individual's frantic effort to avoid abandonment. Affect and mood disturbances are outlined in Criterion 6 and 8.

Criterion 6 is "affective instability due to a marked reactivity of mood" lasting usually "a few hours and only rarely more than a few days" (APA, 2013, p. 663). Common moods include anxiety, irritability, panic, or despair and are "rarely relieved by periods of well-being or satisfaction" (APA, 2013, p. 664). These mood episodes often reflect "the individual's extreme reactivity to interpersonal stresses" (APA, 2013, p. 664). In addition to instability in affect, persons with BPD may display "inappropriate, intense anger or difficulty controlling anger" (Criterion 8) (APA, 2013, p. 664). These displays are usually followed by guilt and shame and further contribute to the internal feeling of being evil. They are often prompted by situations where caregivers or lovers are seen as "neglectful, withholding, uncaring or abandoning" (APA, 2013, p. 664). Lastly, in response to real or imagined abandonment, "transient paranoid ideation or dissociative symptoms" may occur but are "generally of insufficient severity or duration to warrant an additional diagnosis" (Criterion 9) (APA, 2013, p. 664). These symptoms may be relieved by the "real or perceived return of the caregiver's nurturance" (APA, 2013, p. 664). Again, only five of the aforementioned criteria are required to elicit a diagnosis of BPD.

Feminist Theory

It has been 95 years since women fought for and won the right to vote in the United States of America, and 52 years since Betty Friedan published her groundbreaking novel, *The Feminine Mystique*, promoting feminist ideology. Today, many feminists believe that women are still fighting a male-dominated society that strives to control and oppress its women. We live in a patriarchal society that is "male dominated, male identified, and male centered" (Robbins, Chatterjee, and Canda, 2011, p. 114). This is evident in all facets of society as men have a monopoly on most political, economic, and legal systems in the USA. Men occupy the majority of corporate leadership positions and "women constitute only 2 percent of chief executive officers (CEOs), 14 percent of top executives, and 16 percent of directors at Fortune 500 companies" (Shin, 2012, p. 258). Feminist theory works to counter the norm by bringing to light the injustice of male dominance and female subordination with the end goal of achieving equality and opportunity for women. To achieve this goal, feminist theory works to highlight the notion that women and men are equal, and women should be able to define themselves instead of being defined by a male dominated society.

When asked to define the term feminist, "many people offer up bleak and trite descriptors like angry, shrill, or man-haters" (Robbins et al., 2011, p. 113). Contrary to those stereotypical beliefs, feminist theory is not based on hating men, but rather drawing attention to the inequality women experience at the expense of men. Feminists are not asking for women to be named the superior gender, but that society recognize the "existence of multiple experienced realities, based in different vantage points, and supports women in the formation of their own self-understandings and life aspirations" (Robbins et al., 2011, p. 108). For many, this is a radical notion to accept. Not necessarily because they are against it, but because they live in a world where gender inequality is so interwoven into the fabric of our society it is sometimes hard to recognize the many ways women are oppressed. Subconscious beliefs regarding gender roles are most dangerous to the feminist theory agenda because it can be difficult to alter a belief system not recognized for its disparaging effects. Immediately after the birth or adoption of a baby, one of the first questions asked is whether it's a baby boy or girl. The question seems innocent, but the answer carries heavy weight because it determines "the child's name, gifts given, clothing worn, and how family and friends will interact with the child" (Robbins et al., 2011, p. 113). Acknowledging the importance of gender helps point out what gender means to our society and the consequences of being female or male.

Gender schemas are "mental constructs we hold, often implicit or unconscious, about an individual, group of people, or events" (Robbins et al., 2011, p. 113). Gender schemas are so ingrained that we may not even be aware they exist within our beliefs. They can be accurate or inaccurate, positive or negative. The issue is they oversimplify the characteristics of the group. They are rarely challenged or changed, and when we are confronted a person who defies his or her gender schema, we see him or her as an exception rather than acknowledge that society's accepted belief of their presumed schema is incorrect. If you combine this idea of unconscious gender schemas with our patriarchal society, you can see that women are given a subservient, predetermined gender role with labels such as emotional, unstable, and frantic. Feminist theory takes on the difficult position of making society's unconscious beliefs conscious, then challenging them and reforming them into fair and just beliefs. With a rich history of challenging medical practices and psychiatric science, feminists are fiercely critical of the ways "the machinery of psychiatric diagnosis and treatment has been used to obscure or amplify the psychological effects of patriarchies" (Swartz, 2013, p. 41). Over the last 50 years, feminists have "relentlessly challenged many psychiatric diagnoses" (Marecek & Gavey, 2013, p. 3). BPD is one of the diagnoses targeted in the feminist debate over gender oppression in psychiatric practice.

A Feminist Critique of Borderline Personality Disorder

Feminists have long been critics of psychological theories and psychiatric diagnoses and claim much of the literature administered today is "the ideology of a masculinist society dressed up as objective truth" (Marecek & Gavey, 2013, p. 3). The diagnostic criteria, or overall existence, of BPD is a topic especially criticized by prominent feminists who claim it to be the "worst of psychiatry as a mechanism of regulatory control that is historically and culturally predisposed to find women defective and sick" (Lester, 2013, p. 71). The critique of BPD is centered on the idea that BPD is a means to discriminate and marginalize women, deeming their emotions as pathological and their behavior dysfunctional. BPD is seen as "the most pejorative of personality labels which is little more than a shorthand for a difficult, angry, female client" (Ussher, 2013, p. 66). Given the scholarly literature and professional research of today, there is vast support indicating a gender related issue with BPD. This is evidenced by the fact that "BPD is diagnosed predominantly (about 75%) in females" (APA, 2013, p. 666). However, the belief that the issue lies in the possible gender bias of the diagnostic criteria can be challenged and instead, it can be argued the issue is in the ambiguity of the diagnostic criteria altogether. The vagueness of the diagnostic criteria makes it "easily extendable to anyone doctors find inscrutable, provocative, or even merely annoying" (Lester, 2013, p. 71). The general diagnostic criteria paired with our patriarchal society leads to a disproportionate number of women being diagnosed with BPD.

Emotionality, dependency, sensitivity, and instability are characteristics associated with BPD. They are also characteristics "consistent with historically feminine dispositions" (Lester, 2013, p. 71). It cannot be that this personality disorder is solely evidenced in women, so why are so many of the characteristics referenced associated with traditional feminine qualities? Maybe the issue is not in the emotional characteristics identified, but rather in the way those emotional characteristics are viewed differently in genders. Because our society is patriarchal in nature, male gender-role normative behavior is often seen as the baseline for evaluation of illness and health. Impulsive behavior with casual sexual activity (Criterion 4) is "often socially normalized if not valorized" in men, but is pathological in women (Lester, 2013, p. 71). The exact same behaviors, emotions, and characteristics seen in a male as normal, may be viewed as pathological for a woman. In addition to gender-based qualifiers to determine whether certain behavior is pathological versus characteristic, BPD has criteria that could be considered unfair to both men and women. Using the terms "frantic" and "inappropriate" as qualifiers entail judgments about "appropriate ways to respond to abandonment or express anger" (Lester, 2013, p. 71). Again, the criteria for BPD are arbitrarily dictating when certain behavior is appropriate when displayed in what gender.

It is arguably "a go-to diagnosis within a medical system that disproportionately interprets strong emotions in women as symptoms of pathology" (Lester, 2013, p. 71). This double standard of assessment is especially dangerous when diagnosing personality disorders. Gender roles constitute a series of "core and enduring personality traits" acquired through social norms and customs (Brown and Ballou, 1991, p. 212). The DSM defines a personality disorder as "an enduring pattern of inner experience and behavior that deviates markedly from the expectations of the individual's culture" (APA, 2013, p. 645). By that logic, a personality disorder could be viewed as parallel to a gender role. If our society says women should be "polite and compliant," a disorder with criteria outlining the exact opposite of said behavior could be considered to be a women's disorder (Ussher, 2013, p. 64). The interpretation of behaviors and emotions is where the gender issue comes into play during the assessment and diagnosis of a person with BPD.

When interpreting behaviors and emotions, it would be difficult, if not impossible, to make judgment calls without being affected by cultural and societal constructs. It is unrealistic to think a clinician can make a completely unbiased assessment of a human because individuals are "inevitably, affected by the group the person belongs to and our socially ingrained, unconscious schemas" (Robbins et al., 2011, p. 114). The dominant ideologies of our current society state "women express emotion because they are emotional creatures, but men express emotion because the situation warrants it" (Barrett & Bliss-Moreau, 2009, p. 654). If society understands a woman's expression of emotion as an internal character flaw rather than a reaction to an event, it would make sense that more women are given the diagnosis of BPD because the essence of a personality disorder is "an enduring pattern of inner experience and behavior" (APA, 2013, p. 645). The lack of clear distinction between a characteristic and dysfunctional character pathology leaves "ample room for the individual clinician with his or her pre-existing biases to interpret certain personality traits as disordered, while exempting other traits, perhaps equally as dysfunctional, from being described as a form of psychopathology" (Brown & Ballou, 1992, p. 210). The ambiguity in the diagnostic criteria for BPD contributes to the issue of gender-biased interpretation because it leaves a gap in the assessment that could be filled with the clinician's bias based on societal constructs.

Conclusion

Having more specific diagnostic criteria would counteract the possibility of gender bias in assessment and diagnosis. It would also help to add descriptive words that play more towards masculine characteristics. Making these changes would greatly help clinicians avoid the pitfalls of interpreting behaviors as pathology in women and normative in men. Instability in interpersonal skills is too broad and could look different from person to person. The broadness combined with descriptors associated with feminine characteristics lead to overdiagnosis in women and underdiagnosis in men.

The diagnostic criteria for BPD are detrimentally vague, rather than gender-biased. It is understandable how feminists see BPD as an attack on women and a means to "classify women's sexuality and experience as anxiety and depression" because it is given mostly to women (Swartz, 2013, p. 41). Some believe the best way to counterattack gender discrimination is to do away with the disorder entirely, much in the same way Hysteria and Homosexuality were eventually removed from the DSM (Ussher, 2013, p. 64). To completely do away with this disorder would be a disservice to those who may be suffering with BPD and need medical attention because "diagnosis may be helpful; or conversely the absence of such a diagnosis may at times be harmful" to those with psychological difficulties (Swartz, 2013, p. 46). BPD is a genuine disorder that should not be tossed aside because of its issue with female prevalence. Instead, the criteria should be reexamined and further specified. Having more specific criteria would narrow the opportunity for a gender bias to affect an assessment. Also, adding characteristics associated with male gender roles may guide the clinician towards a more thoughtful and competent diagnosis of BPD in men and women.

References

American Psychiatric Association (2013). Diagnostic and statistical manual of mental disorders (5th ed.). Washington, D.C: American Psychiatric Association.

Barett, L. F., & Bliss-Moreau, E. (2009). She's emotional. He's having a bad day: Attributional explanations for emotional stereotypes. Emotion, 9(5), 649-658.

Kernberg, O. F., & Michels, R., (2009). Borderline Personality Disorder. AM J Psychiatry, 166 (5), 505-508.

Lester, R. (2013). Lessons from the borderline: Anthropology, psychiatry, and the risks of being human. Feminism and Psychology, 23(1), 70-77.

Marecek, J., & Gavey, N. (2013). DSM 5 and beyond: A critical feminist engagement with psychodiagnosis. Feminism and Psychology, 23(1), 3-9.

Robbins, S. P., Chatterjee, P., & Canda, E. R. (2011). Contemporary human behavior theory: A critical perspective for social work (3rd ed.). Upper Saddle River, NJ: Pearson Education.

Shin, T. (2012). The gender gap in executive compensation: The role of female directors and chief executive officers. The Annals of the American Academy of Political and Social Science. 639(1), 258-278.

Swartz, S. (2013). Feminism and psychiatric diagnosis: Reflections of a feminist practitioner. Feminism and Psychology, 23(1), 41-48.

Ussher, J. M. (2013). Diagnosing difficult women and pathologising feminity: Gender bias in psychiatric nosology. Feminism and Psychology, 23(1), 63-69.

Conduct Disorder, Race, and The Environment

Rachel Cope

Some will argue that we are living in a post-racist world. Others say that racism is still very much alive. Examining recent news stories about Michael Brown and Tamir Rice, just to name few, we see young unarmed black youth being gunned down by white police officers. When the Department of Justice released its report on the policing practices in Ferguson, Missouri, in response to the shooting of Michael Brown, it explicitly states, "Ferguson's approach to law enforcement both reflects and reinforces racial bias, including stereotyping" (Investigation of the Ferguson Police Department, 2015, pg. 4). Additionally, a study conducted in Iowa and Georgia between the years 1997 and 2002 found that over 80% of African-American youth had experienced some kind of racial discrimination before the age of thirteen (Martin, M., McCarthy, B., Conger, R., Gibbons, F., Simons, R., & Cutrona, C., 2010).

Not only does individual racism exist (racist comments and insults) but there is also a system of institutional racism embedded in the fabric of American society. Using Critical Race Theory as a lens through which to view the mental illness conduct disorder (CD), I will discuss how individual and institutional racism simultaneously oppress youth of color and deteriorate their mental health. Minority youth are more often exposed to the risk factors associated with CD such as poverty, community violence, and neighborhood disorganization. Additionally, youth diagnosed with CD tend to exhibit antisocial behaviors. Those behaviors can lead to incarceration in adulthood thus making CD a stigmatizing label for a young person. In this paper, I will also argue for a more holistic approach in the mental health field when working with minority youth exhibiting externalizing behaviors that meet criteria for CD.

Conduct Disorder

The Diagnostic and Statistical Manual of Mental Disorders (5th ed; American Psychiatric Association, 2013), defines a mental disorder as a syndrome that disturbs an individual's cognition, emotion regulation, or behavior. This disturbance must create significant distress in the individual's social or occupational life. The DSM 5 continues, "conflicts that are primarily between the individual and society are not mental disorders *unless* the deviance or conflict results from a **dysfunction** in the individual" (italics and bold mine; APA, 2013, pg. 20). It is certainly correct to say that the mental illness conduct disorder disturbs an individual's behavior and ability to regulate his/her emotions, but is this disturbance caused from an internal dysfunction or is the disturbance formed in response to a negative environment? This question will be discussed in more detail below.

The diagnostic criteria for CD represent a set of externalizing behaviors that cause significant impairment in social, academic, or occupational functioning. An individual must meet three of the following fifteen criteria in the past twelve months, with at least one criterion present in the past six months: aggression to people and animals (includes bullying, physical fights, cruelty to animals and people), destruction of property, deceitfulness or theft (breaking and entering, lying, stealing), and lastly, serious violation of rules (staying out past curfew, truancy, and/or running away from home) (APA, 2013). Jerome Wakefield and colleagues (2002) mention that the criteria in the DSM 4 make no reference to an exclusion related to a negative environment that would allow for a classification of CD not caused by internal dysfunction. The DSM 5 fairs no better. While there are specifiers for age of onset, "with limited prosocial emotions" (pg. 470) including lack of remorse or guilt, callousness, unconcerned about performance, and shallow or deficient affect as well as severity of symptoms (APA, 2013), there is no recognition in the DSM 5 that these behaviors could be natural responses to a negative environment.

One study tried to determine how clinicians would respond to the internal dysfunction/negative environment conundrum. They tested two hypotheses on a group of sixty-two psychology and fifty-five social work graduate students; hypothesis 1) clinicians would judge environmental-reactive antisocial behaviors as non-disorders even when the behaviors meet criteria for CD, and hypothesis 2) clinicians would judge antisocial behavior brought on by internal dysfunction as disorders (Wakefield, J., Pottick, K., & Kirk, S., 2002). The results were as they expected. The student clinicians overwhelmingly judged in favor of internal dysfunction determining whether a client is labeled with a mental disorder.

This line of thinking follows what the DSM 5 states as a mental illness. The study also concluded that the student clinicians believed youth displaying antisocial behavior, regardless of the reason why, were in need of professional help or treatment (Wakefield, J., et al., 2002). These results are from a designed experiment using judgments from Masters level students with only an average of four years clinical work experience. In real-life clinical settings, determining if an adolescent is exhibiting antisocial behaviors due to internal dysfunction as the DSM 5 defines a mental illness or the behaviors are in response to a negative environment is an extremely difficult task. The clinician has a grave responsibility to give the correct diagnosis as it will determine the treatment plan as well as how the youth internalizes his/her issues.

Differential Diagnoses and Race

African-American and Latino youth are disproportionately given the diagnosis of CD compared to their white counterparts (Mizock, L., & Harkins, D., 2011). Often clinicians and the public-at-large view CD as a more stigmatizing label compared to other diagnoses such as attention deficit/hyperactivity disorder (ADHD) or attachment disorder (Mizock, L., et al., 2011). CD is often characterized by antisocial behavior and is a prerequisite to antisocial personality disorder (Mizock, L., et al., 2011). Antisocial behaviors tend to lead to a life of criminality and incarceration, thus explaining why this is such a stigmatizing diagnosis. Clinicians must examine all factors in a client's life that may contribute to aggressive or antisocial behavior before labeling a client with the diagnosis CD.

Research shows that antisocial behaviors that meet criteria for CD are accompanied by feelings of depression, separation anxiety, and attachment disorders (Mizock, L. et al., 2011). Many youth express conduct problems in response to these underlying feelings. Additionally, youth may seek out illegal substances to cope with said feelings resulting in additional behavior problems (Mizock, L., et al., 2011). Mizock and Harkins (2011) comment that the research shows, "clinicians tend to misperceive conduct disorder in adolescents of color who present with behavior symptoms of other psychological disorders" (pp. 245). Youth who engage in destructive behavior and/or show strong opposition to rigid rules in the home or class-room could have other pathologies such as bipolar disorder, ADHD, or obsessive-compulsive disorder (OCD) (Mizock, L., et al., 2011). These other diagnoses are often overlooked when an African-American or Latino adolescent presents with aggressive/violent behaviors.

The same holds true when looking at the diagnosis of autism spectrum disorder in white and African-American children. Mandell, Ittenbach, Levy, and Pinto-Martin (2006) conducted a study using a sample of families receiving Medicaid in Philadelphia from July 1993 through June 1999. They found that African-American youth were diagnosed with autism spectrum disorder on average 1.4 years later than their white counterparts and spent eight more months in mental health treatment before being diagnosed (Mandell, D., et al., 2006). Additionally, African-American children who were eventually given the diagnosis of autism spectrum disorder were three times more likely to receive another diagnosis first, usually being that of CD or an attachment disorder (Mandell, D., et al., 2006).

The question is why. Mandell and colleagues give several important reasons. They state that child presentation of the illness, the parents' behavior in response to symptoms, and parents' recognition and interpretation of the symptoms all play an important role in determining how a clinician gives a diagnosis (Mandell, et al., 2006). Discrimination is present as well; "the more frequent diagnosis of conduct disorder among African-Americans may be associated with clinicians erroneous beliefs regarding the increased frequency of conduct disorder among African-American children" (Mandell, D., et al., 2006, pg. 1800). If clinicians preemptively assume that African-American children presenting with aggressive behaviors have CD over another diagnosis like ADHD or autism spectrum disorder then the mental health profession has long way to go in dismantling racism within.

We find similar results in residential treatment facilities. Cameron and Guterman (2007) discovered that white boys and girls were given the CD diagnosis least often but were equally as often or more often in clinical range on the CBCL aggressive scale compared with their African-American and Latino counterparts. These results came from a study of 1,173 youth in residential group care, group homes, or therapeutic foster care settings. This raises the question: Why are African-American and Latino youth consistently over-diagnosed with the stigmatizing mental illness CD when their white peers are exhibiting the same behavior but often receive a different diagnosis? Additionally, it could be argued that the diagnosis of CD becomes a self-fulfilling prophecy for youth who are already a part of the disadvantaged minority. Critical Race Theory claims that, "society is fundamentally racially stratified and unequal, where power processes systematically disenfranchise racially oppressed people" (Hylton, K., 2012, pg. 24). I posit that by disproportionately labeling minority youth with the mental illness conduct disorder, the mental health field is one of those power processes systematically oppressing youth of color.

Risk Factors

There are both individual risk factors as well as environmental concerns to consider when examining the diagnosis of conduct disorder. Because of racial stratification in society, minority youth are more often exposed to some of these risk factors. Both community disorganization and community violence are linked to conduct problems in adolescents (Hill, J., 2002). One study reported that over 65% of African-American and Latino youth in an inner city neighborhood of Chicago witnessed some kind of violence within the past year (Hill, J, 2002). This violence included shootings, stabbings, and other types of street violence.

Additional studies by Slattery and Meyers (2013) found that exposure to community violence was the "most powerful and pervasive" (pg. 53) predictor of antisocial behavior in adolescents, even when controlling for interpersonal and demographic risk factors. It is imperative for clinicians to consider the community factors that are affecting African-American and Latino youth when deciding whether or not to give a diagnosis of CD. The reality of racial stratification in American society has put African-American and Latino populations at the bottom of the social and political totem pole. This critically impacts their quality of life and life chances (Brown, T., 2003), which can lead to mental health issues.

Other risk factors for CD include individual temperament, low socioeconomic status, verbal deficits/low IQ, and perinatal complications. Irritable babies may evoke hostility in their mothers, which can in turn cause antisocial behaviors in children (Hill, J., 2002). Additionally in males a combination of lack of self-control and the number of changes in parental figures before the age of thirteen was a solid predictor of conviction of a violent offense (antisocial behavior) by the age of eighteen (Hill, J., 2002). Other more obvious factors such as childhood abuse contribute to aggression in children.

Verbal deficits also prove to have a significant impact on whether an adolescent might meet criteria for CD. Hill (2002) claims that children with conduct problems show greater deficits in language-based verbal skills. For example, children who are unable to assert themselves using verbal skills may resort to aggression to gain control in specific social exchanges (Hill, J., 2002). Other children might show increased aggression when they do not possess the language skills to accurately express their emotions and the critical thinking abilities to understand the emotional state of others (Hill, J., 2002). Aggression/irritability is a rather logical response to feelings of inadequacy in both language skills and critical thinking. Hill (2002) lists several more risk factors associated with CD which include: problems with behavioral activation and inhibition systems, lack of proper executive functioning, poor peer relationships, attachment issues, inconsistent parenting practices, and perinatal complications. The latter includes obstetric complications such as poor antenatal care, alcohol and drug use during pregnancy, and poor socioeconomic conditions (Hill, J., 2002).

Hill (2002) concludes by commenting that youth violence tends to be more prevalent in neighborhoods where drugs are readily available and adolescents are exposed to adult criminality and racial prejudice and where there is a perceived lack of community organization. As discussed above, violent or aggressive African-American and Latino youth are more commonly given the diagnosis of CD compared to their white counterparts who receive less stigmatizing diagnoses such as ADHD or attachment disorder. Additionally, minority youth are more often exposed to the environmental risk factors associated with CD due to racial stratification in American society.

Enduring Racism: Effects on Mental Health

Does individual and structural racism affect the mental health of African-Americans in the United States? Williams and Williams-Morris (2000) in their paper "Racism and Mental Health: The African American Experience" state that institutional racism restricts socioeconomic mobility thus forcing people to live in poor conditions in disorganized communities. Some scholars call this "racism without the racists" (Bonilla-Silva, 2006; Ford, 2008 as cited in Martin, M., et al., 2010, pg. 663).

This type of racism restricts African-Americans' access to adequate housing, employment opportunities, and other neighborhood resources (Martin, M., et al., 2010) thus lending itself to the formation of inner cities with high crime rates. This social isolation from normative society creates a cognitive landscape where committing crimes is an adequate means for addressing the frustration that stems from perceived inequality (Williams, D., et al., 2000). It is also a way to gain respect. Due to racial segregation and economic disadvantage, minority youth struggle to find status in arenas such as school or work; turning to the streets is often a way to get the desired respect that all youth yearn to receive (Martin, M., et al., 2010).

Additionally, personal discrimination as well as accepting one's inferiority as part of a minority group has deleterious affects on psychological functioning (Williams, D., et al., 2000). A strong sense of identity and self-worth are protective factors needed to ward off mental illness. When discrimination becomes internalized such that marginalized racial populations accept the negative societal beliefs and stereotypes about themselves, this can lead an individual to feel worthless and powerless (Williams, D., et al., 2000). In a study examining delinquency in black youth it was determined that discrimination was significantly related to violent as well as general delinquency (Martin, M., et al., 2010). Negative treatment like racial discrimination can generate negative emotions such as anger, frustration, and resentment, which can lead to externalizing behaviors like shoplifting and vandalism to exact revenge (Martin, M., et al., 2010).

African-American youth are overly exposed to external risk factors linked to developing CD and are more often given the diagnosis of CD compared to their white counterparts who exhibit similar behavior. They also may develop internal feelings of worthlessness as a result of racism

(both institutional and individual) in American society. These factors combined have harmful affects on the emotional wellbeing of an already disad-
disadvantaged population.

Conclusion

It is vital to consider who determines what is mental health and
what is mental illness in a society. When the dominant cultural group determines this boundary, it must be carefully examined. In the 19th Century,
two psychiatrists claimed that black slaves suffered from a mental illness
called "drapetomania" – a psychiatric illness indicated by an *externalizing
behavior* of black people wanting to run away from slavery (Brown, T.,
2003, pg. 299). In another 150 years, will conduct disorder be the "drapetomania" of the 21st Century?

Clinicians have a grave responsibility to holistically treat every client
that walks through their doors. Minority youth exhibiting behaviors that
warrant a diagnosis of CD must be holistically assessed to determine the
deeper meaning behind their behaviors. Other pathologies such as ADHD
or autism spectrum disorder should be discussed, as well. Structural change
in American society is needed so that minority youth are less exposed to the
risk factors associated with CD and therefore have a greater chance at
achieving stable mental health in their lifetime.

References

American Psychiatric Association. (2013). *Diagnostic and Statistical Manual of Mental
Disorders* (5th ed.). Washington DC: American Psychiatric Association.

Brown, T. (2003). Critical Race Theory Speaks to The Sociology of Mental Health:
Mental Health Problems Produced by Racial Stratification. *Journal of Health And
Social Behavior, 44:3*, 292-301. Retrieved on 28 March 2015 at
http://www.jstor.org/stable/1519780

Cameron, M. & Guterman, N. (2007). Diagnosing Conduct Problems of Children and
Adolescents in Residential Treatment. *Child Youth Care Forum, 36*, 1-10. DOI:
10.1007/s10566-006-9027-6

Hill, Jonathan. (2002). Biological, Psychological, and Social Processes in the Conduct
Disorders. *Journal of Child Psychology and Psychiatry, 43:1*, 133-164. DOI:
10.1111/1469-7610.00007.

Mandell, D., Ittenback, R., Levy, S., & Pinto-Martin, J. (2007). Disparities in Diagnosis Received Prior to a Diagnosis of Autism Spectrum Disorder. *Journal of Autism Developmental Disorder, 37*, 1795-1802. DOI: 10.1007/s10803-006-0314-8

Martin, M., McCarthy, B., Conger, R., Gibbons, F., Simons, R., Cutrona, C., & Brody, G. (2010). The Enduring Significance of Racism: Discrimination and Delinquency Among Black American Youth. *Journal of Research on Adolescence, 21:3,*662-676. DOI: 10.1111/j.1532-7795.2010.00699.x

Mizock, L. & Harkins, D. (2011). Diagnostic Bias and Conduct Disorder: Improving Culturally Sensitive Diagnosis. *Child and Youth Services, 32:3*, 243-253. DOI:10.1080/0145935X.2011.605315

Slattery, T. & Meyers, S. (2013). Contextual Predictors of Adolescent Antisocial Behavior: The Developmental Influences of Family, Peer, and Neighborhood Factors. *Child Adolescent Social Work Journal, 31*, 39-59.DOI 10.1007/s10560-013-0309-1

United States Department of Justice Civil Rights Division. (2015). Investigation of The Ferguson Police Department. Retrieved on 28 March 2015 at http://www.justice.gov/sites/default/files/opa/pressreleases/attachments/2015/03/04/ferguson_police_department_report.pdf

Wakefield, J., Pottick, K., Kirk, S. (2002). Should the DSM-IV Diagnostic Criteria for Conduct Disorder Consider Social Context? *American Journal of Psychiatry, 159:3*, 380-386. http://dx.doi.org/10.1176/appi.ajp.159.3.380

Williams, R. & Williams-Morris, R. (2000). Racism and Mental Health: The African-American Experience. *Ethnicity and Health, 5:3/4*, 243-68. DOI: 10.1080/135578500200009356

Body Dysmorphic Disorder Through the Lens of Feminist Theory

Mackenzie Coen

Introduction

With unrealistic body image representation in today's culture it is important to understand how and why men and women are experiencing Body Dysmorphic Disorder. Looking through a feminist theoretical lens, allows us to see how BDD may be a result of unconscious manifestations of feelings of inadequacy. Also it is important to differentiate how men and women experience BDD by looking at the values that men and women place on themselves. By dissecting how men and women experience BDD we can begin to understand how feminist theory favors neither men nor women when it is addressing body image and self worth.

Body Dysmorphic Disorder

According to the DSM 5, Body Dysmorphic Disorder is classified as an Obsessive Compulsive Disorder. The criteria include: preoccupation with flaws not perceivable or observable to others, repetition of mirror checking, grooming, or other obsessive and compulsive reassurance behaviors. It also includes preoccupation that causes distress or impairment in important areas of functioning and the appearance concerns cannot be better explained by concerns with body fat or weight (whose symptoms meet those for eating disorder criteria, see American Psychiatric Association, 2013. The interesting aspect of this disorder is that it is not classified as an eating or elimination disorder by definition. In fact, it cannot be classified as BDD if the obsessions can be explained by an eating disorder. Someone who suffers with BDD will not be obsessed with losing weight but instead have an obsession with the way their body appears. Therefore, someone who is suffering from this disorder is significantly impaired by the belief that his or her body does not look or appear a certain way.

Even though this disorder often causes severe impairment, it is often missed in clinical settings (Phillips, 2004). The areas that a person suffering with BDD could obsess about are anywhere from facial features, to skin tone and bodily features. In a clinical setting, the therapist treating this type of client may see them continually putting on make-up, fixing their hair, or even skin picking. When it comes to BDD, nearly as many men suffer as women. However, more women are likely to seek treatment than men, leading to more documented cases of women with BDD (Phillips & Castle, 2001). Often it is seen that women are more obsessed with appearance than men; however, men have been shown to have obsessions with hairline and muscle tone during clinical sessions (Neziroglu, Slavin, n.d). Given what can be said about BDD from practice and the DSM 5, it can be seen that this disorder affects both men and women. However, can using a feminist theory perspective in looking at BDD affect the way a diagnosis may be made? In addition, feminist theory may show that both men and women are vulnerable to BDD.

Feminist Theory

Feminist theory has become a popular topic but has been around a lot longer than many would expect. The first wave of feminist theorists came around the time of the suffragette movement that forced laws giving women the right to vote in 1919. The second wave made its presence again between the 1960's to the 1980's. The third wave came around the 1990's, as the second wave did not carry lasting benefits for women ("History and Theory of Feminism", n.d). Today, our culture's opinions of feminist theory have drastically changed. While feminism still carries the beliefs that started the movement, modern society has created a new image for feminists today in the United States.

The working concept of feminist theory is that women and men should be equal, politically, economically and socially. While this working definition is what drives the theory, it is important to understand how feminist theory works in social work practice in regards to how it affects men and women. The theory itself has used with therapeutic approaches made by many social workers in the field. Social workers using feminist theory strive to improve women's mental well being by addressing the linkage of private troubles to current social positions and/or status. On the other hand, a social worker using feminist theory with men in practice requires a reconceptualization of masculinity within feminist insights (Dominelli, 2003).

This theory also proposes that there is a root cause of women's problems that is correlated to women's social positions and roles. These roles have also been a product of gender socialization, which is also an important aspect to consider when examining feminist theory. However, what it fails to be mentioned in research regarding feminist theory in practice is how it affects men as well The typical schema that is known as gender socialization is that boys and girls are raised to be two completely different genders. The feminist psychoanalytic perspective states that gender is embedded in the unconscious of each and every being (Lorber,1994). However, it may not be strictly gender roles and socialization that drive the core values of feminist theory. I will use this theory and gender socialization as a means to further explore how it may affect diagnosis in the field, specifically Body Dysmorphic Disorder.

Gender Roles on Socialization Values

Previously, when discussing feminist theory the idea of gender socialization was touched upon. When looking at gender socialization, especially in regards to feminist theory it is important to open up the working definition that might already exist. For the purpose of analyzing the theory and BDD I chose to look beyond the typical "boys play with trucks and girls play with dolls" passé stereotypes. Instead I chose to look at how young girls and boys classify their self worth.

Your self worth is how you perceive what you need to have value in the world. Males are often taught to have value irrespective of their looks, but instead are taught to place value on their ability to perform or be dominant. While females, even the most successful and intelligent, may still feel the requirement to look a certain way to be valuable to the world. However, we need to see how these values that are placed on gender may negatively affect both sexes. Looking at this from a feminist theory we could see that females have been oppressed because if they are seen as overly aggressive they will be viewed in a negative light. Furthermore, it can be assumed that the pressure to fulfill a "woman's role" according to society puts a glass ceiling on a woman's self worth.

On the other hand, men have been socialized to put such a high value on the need to perform, that the failure to perform may result in feelings of inadequacy. Looked at this from another point of view, one could argue that women have it "easy" and that a value of looks is simpler than being expected to perform and be dominantly aggressive. In order to understand BDD in regards to feminist theory it is important to consider both angles that one can look at in terms of gender value.

Linkage of BDD and Unconscious Feelings of Inadequacy

When considering BDD and feminist theory, unconscious manifestations must be addressed. As stated previously, social workers using feminist theory work under the belief that their female clients' feelings of inadequacy have manifested as skewed image problems that lead to BDD. Women's body image is a hot topic of today's culture in the United States. However, there have been cases of both men and women who suffer from BDD. Therefore it is crucial to understand that feelings of inadequacy can be manifested in both men and women. While our culture may favor women's struggle in physical appearance, by doing so men's body image is being neglected in the process. It all goes back to what makes men and women feel like they have value in the world. Men are stereotypically projected to need to be strong and assert dominance in physical appearance.

In tying this all back to feminist theory we have to look at the main foundation feminist theory is built on. This is the idea of equality between the sexes. Therefore, what needs to be stressed is that in order to follow this framework of equality, the lens of feminist theory should actually encompass men as well. When talking about subconscious manifestations of inequality and oppression of females, it needs to be taken into careful consideration that by doing so we actually might be boxing the genders in. If we solely focus on the manifestations of inequality being a contributor of BDD, we are saying that females are the only ones who would struggle with body image.

To strengthen the importance of men's image body issues being an equal concern to women's, consider a research study done on eating disorders between men and women by Striegel, Bedrosian, Wang, & Schwartz (2012), which compared men and women who reported binge eating. While the data stated that women had a slightly higher prevalence of binge eating, the research found some interesting results which suggested that binge eating in men was associated with significant impairment. Furthermore, they hypothesized from their research that men may be more reluctant to seek treatment for body image and eating disorder struggles. They noted that the researchers considered the fact that body image and eating disorders are seen as a "female problem". Even though this study is showing an example with eating disorders, we can deduce that men are reporting less feelings of inadequacy of self-image. It is clear that data can show men and women are both struggling with BDD and how they value their self-image.

How Men and Women Experience BDD

The next area to address to make sense of BDD through a feminist theoretical lens would be to delve into how BDD may look in men versus women. While men and women are both diagnosed, it is more likely that women will seek treatment. However, this may be because BDD looks differently for men than it does for women. Men are often preoccupied with thinning hairline, muscle tone and the appearance of their genitals (Phillips & Castle, 2001). In fact a sub disorder of BDD called Muscle Dysmorphia applies almost exclusively to men. Men may exhibit complaints of feeling "too small" compared to other men, whereas women often complain of feeling "too big" in comparison to other women. So it is obvious that males in fact have unrealistic expectations of their bodies just like females. Another factor to contribute to men not reporting concern about body image is that they do not see this dissatisfaction or obsession with their body as a disorder. Looking at this disorder back in the lens of feminist theory, it is important to consider how this theoretical perspective might be affecting men.

Feminist Cultural Oppression of
Body Image Standards

So far we have looked at how women suffer with BDD versus men is different, but paradoxically the same. According to DSM 5 criteria BDD falls into the Obsessive Compulsive category. However, like many other disorders it can be co-occurring with other disorders. Just as feminist theory states that women's oppression may be the underlying issue that leads to BDD, this same theory can look at men's mental health as an oppressed state. Here we should look at the difference between the two oppressed states. Women may experience oppression by feeling like they have been limited and held to different standards than men. Men may feel this oppression by being boxed into this image of what it means to be a man. To many this may not seem like oppression, but again by definition of feminist theory, it is important to look at men and women equally. Men and women are very different, but they are both equally are at risk of oppression in different manners. For example, a scrawnier man might be looked down on by other men and even women. This oppression may put him at a higher risk for BDD, depression, anxiety, and other related disorders. The muting of feelings of failure or dissatisfaction makes it harder for clinicians to diagnose BDD and such disorders in men (Rabinowitz & Cochran, 2008). Therefore we know that men are holding these oppressed feelings but may not exhibit them like women.

BDD and the Aging Female Body

While this paper addresses men and BDD according to this theory, it does discount the women who are suffering with BDD as well. The seriousness of BDD in both genders is important to emphasize. The important idea is that feminist theory is not limited in effects to only women. However, women are clearly suffering with BDD as well. An important aspect of BDD for women specifically, is body image as they age. The value on women to be young looking, thin, and wrinkle free is something women struggle with as they get older (McCormick, 2008). A clinician using feminist theory may look at this as women's dissatisfaction due to expectations put upon women.

These expectations could be for example to have children. While it is natural that women's bodies change over time, women see this change as negative. Under this theory the pressure to have children by society could result in a lack of satisfaction of body image as women age. While men's bodies change with age as well, under this theory women may have more cultural pressures that affect how they view their aging bodies. In addition, men are able to reproduce most of their lives, whereas women cannot have children after menopause. Therefore while men may show dissatisfaction with their bodies, as they get older, they may still hold value in their minds compared to women who are no longer able to have children. The way women and men age visually is much different, so it is common for BDD rates to go up in women post menopause.

Changes in Cultural Body Image Trends

Another important aspect of body image distortion and appearance in men and women is how it has changed over the years. What is very interesting is that how we look at what is deemed attractive in a male has not changed much over the last several decades. However, the "ideal" woman's image has changed immensely. Starting in the 50's curves were seen as ideal and attractive, take for example Marilyn Monroe. By the 70's and 80's thin lean bodies were ideal for women. In the 90's an attractive women was said to be rail thin with no chest or lack of curves. Now in the 2000's we are seeing a new trend once again. One of these is the emphasis on large chest, tiny waist and wide hips, giving the hourglass illusion. Another body trend for women is positive emphasis on muscle tone throughout the body, which was once solely a masculine feature. A feedback about this trend from feminist theory is that we are now holding the value that both men and women can have muscle tone and be attractive.

Healthy body image is all over the media; however how can these changing physical trends affect BDD? For women, while being strong is seen now as an attractive feature, there are still criticisms as to how much muscle is too much or "manly." The traditional quote that "beauty is in the eye of the beholder" says it all. Consequently, men and women who struggle with BDD see image as a true meaning of value. The changing trends may bring stress on body image to women. On the other hand, men are still facing the same standard of being "tough," muscular and manly. The positive side of this for feminist theory is that the image of strong women equal to men is now being more widely accepted. However, this does not mean that BDD rates will decrease. In fact, a trend that I predict is that the number of women suffering from muscle dysmorphia will go up. It is important to touch on modern beauty trends because they do in fact affect BDD rates, and they also effect how body image changes over time.

Feminist Theory in Practice

The take away from this cannot be concluded without applying BDD through the lens of feminist theory in regards to practice. First, what might a clinician using a feminist theoretical framework focus on? Conceptually the clinician will focus on righting the wrongs done to women historically speaking (Grosz, 2010). Therefore, it is important to ask, how does feminist theory affect men? More specifically, how does feminist theory affect practice with men and women suffering with BDD? Feminist theory is the work of producing new thought beyond patriarchal concepts (Grosz, 2010). That idea in itself can explain how feminist theory can help both men and women. By looking beyond patriarchy and masculine concepts, clinicians can peel back the layers from women and men. When men are trapped in a patriarchal mind set, this means they are also most likely to place value solely on how they look as a "man." The bottom line of this is, feminist theory attempts to break the stereotype of what a "man" looks like and what a "woman" should look like. Therefore, in practice a clinician using this lens to look at BDD will work with the client to help find the subconscious layers of what he or she believes beauty is for their gender. Persons suffering with BDD already have an image in their mind of what their body should look like, but where they get this image from may be something much deeper in the psyche. A clinician treating BDD can use this theory to dissect what the clients thinks is ideal "physically" and help to ultimately build a realistic expectation.

Conclusion

At first glance, we can tell when someone is suffering from BDD by addressing the criteria using the DSM 5. However, in order to apply this to theory and practice, this disorder has to be considered on a multidimensional basis. It would be limiting to only use DSM 5 criteria to diagnose BDD. As previously stated, there are unconscious beliefs that may be the underlying causes of BDD. This is not limited to women from a feminist theoretical framework, but as we can see men are affected as well. If unrealistic standards are held for men and women, this is going to cause an internal struggle. Overall, to begin looking at BDD with this mindset we have to assume that every man or woman suffering from negative body image is the product of values placed on gender roles. When men are expected to be broad, manly and muscular, it is setting up a risk of failure for men. The same idea can be used for women, when there is an expectation for them to fill women's roles, be feminine, petite and polite. The only way to break the cycle of oppression of both genders is to use this theory as a guide. Ultimately feminist theory can allow clinicians to see that BDD has many layers that are affected by the environment, culture, and individuals' perceptions of what true self worth is.

References

American Psychiatric Association. (2013). Diagnostic and statistical manual of mental disorders (5th ed.). Arlington, VA: American Psychiatric Publishing.

Grosz, E. (2010). The Practice of Feminist Theory. *Differences-a Journal of Feminist Cultural Studies*. doi:10.1215/10407391-2009-019

History and theory of feminism. (n.d.). Retrieved from http://www.gender.cawater-info.net/knowledge_base/rubricator/feminism_e.htm

Lorber, J. (1994). *Paradoxes of gender*. New Haven: Yale University Press.

McCormick, M. (2008). Womens bodies aging: Culture, context and social work practice. *Journal of Women and Social Work, 23*(4), 312-323.

Phillips, K. (2004). Body dysmorphic disorder: recognizing and treating imagined ugliness. *World Psychiatry, 3*(1), 12-17.

Phillips, K. A., & Castle, D. J. (2001). Body dysmorphic disorder in men : Psychiatric treatments are usually effective. *British Medical Journal.* doi:10.1136/bmj.323.7320.1015

Rabinowitz, F. E., & Cochran, S. V. (2008). Men and therapy: A case of masked male depression. *Clinical Case Studies.* doi:10.1177/1534650108319917

Striegel, R. H., Bedrosian, R., Wang, C., & Schwartz, S. (2012). Why Men Should be Included in Research on Binge Eating. *British Medical Journal.*

van, W. K. (January 01, 2003). Dominelli, Lena, Feminist Social Work: Theory and Practice. International Social Work, 46, 266.

Anxious Awareness
A Mindful Critique Of Generalized Anxiety Disorder According To The DSM 5

Niki Fox

Introduction

Generalized anxiety disorder is defined by criteria designated by the American Psychiatric Association's (APA) *Diagnostic and Statistical Manual of Mental Disorders, 5th Edition* (DSM 5). While these criteria help clinicians officially diagnose their clients using a checklist of symptoms and generalized conditions, it can also pigeonhole and marginalize an individual's concept of self and understanding of his/her experiences and relationships in life. This paper explores the diagnosis of generalized anxiety disorder (GAD), the required criteria to make this diagnosis, according to the DSM 5, and key components and common experiences reported by individuals diagnosed with this disorder. The concept of mindfulness, as derived from Buddhism, is also defined and examined as a perspective from which to analyze and critique the diagnosis of generalized anxiety disorder and its defining criteria. According to Kabat-Zinn (2013), mindfulness is "the awareness that arises by paying attention on purpose, in the present moment, and non-judgmentally" (p. xxxv). Within the mindfulness perspective, acceptance is understood as "taking each moment as it comes and being with it fully, as it is. We try not to impose our ideas about what we 'should' be feeling or thinking or seeing in our experience" (Kabat-Zinn, 2013, p. 28). Awareness, being in the present, and non-judgmental evaluation, as components of mindfulness and overall acceptance will be used to critique the DSM 5 diagnosis of GAD.

Diagnosis: Generalized Anxiety Disorder

Generalized anxiety disorder (GAD) is defined in part by the DSM 5 as "excessive anxiety and worry (apprehensive expectation) about a number of events or activities" (American Psychiatric Association (APA), 2013, p. 222). Individuals diagnosed with generalized anxiety disorder express "chronic worry," where "simply having anxiety can lead to a strong desire to change the content, intensity, or frequency of internal experiences" (Hayes et al., 1996, as cited by Hayes-Skelton et al, 2013, p. 264). The worry and symptoms experienced in association with generalized anxiety disorder are determined to be pathological as they are persistent, prominent, and cause distress.

Generalized anxiety disorder invades all aspects of an individual's life, occurring without precipitating factors or influences. The anxiety experienced in connection with generalized anxiety disorder persists in intensity disproportionate to perceived actual events or the impacts of the anticipated events. Individuals with generalized anxiety disorder self-report that "they have felt anxious and nervous all of their lives" even though the median onset age is 30 years (APA, 2013, p. 223).

The chronic symptoms of generalized anxiety disorder can wax and wane throughout one's lifespan with low rates of full remission (APA, 2013). The DSM 5 reports that within a 12-month prevalence, 0.9% of adolescents and 2.9% of adults in the general population of the United States have generalized anxiety disorder. Females are twice as likely as males to be identified as having GAD (APA, 2013). There are temperamental risk factors associated with generalized anxiety disorder including inhibited behavior, negative affect (neuroticism) and avoidance of harm or perceived harm.

DSM 5 Criteria For Diagnosis of
Generalized Anxiety Disorder (GAD)

Within generalized anxiety disorder, an individual experiences anxiety that is excessive in combination with worry or "apprehensive expectation,"... "occurring more days than not for at least 6 months, about a number of events or activities (such as work or school performance)" (Criterion A; APA, 2013, p. 222). According to the DSM 5 (2013), there is extreme difficulty controlling the worry and a great deal of cognitive and behavioral effort goes into controlling apprehension (Criterion B). The diagnostic Criterion C specifies "anxiety and worry are associated with three or more of the following six symptoms (with at least some symptoms having been present for more days than not for the past 6 months): 1. Restlessness or feeling keyed up or on edge; 2. Being easily fatigued; 3. Difficulty concentrating or mind going blank; 4. Irritability; 5. Muscle tension; 6. Sleep disturbance (difficulty falling asleep or staying asleep, or restless, unsatisfying sleep)" (APA, 2013, p. 222). It is also noted that, in children, only one symptom in Criterion C needs to be present. According to the DSM 5, the worry, anxiety, and physical symptoms of Criterion C must "cause clinically significant distress or impairment in social, occupational, or other important areas of functioning" (Criterion D) (APA, 2013, p. 222). Generalized anxiety disorder asserts that the physiological effects of a substance or medication or a medical condition do not cause the disorder (Criterion E; APA, 2013). Lastly, the difficulties experienced with GAD are not to be attributed to another mental disorder (Criterion D).

Theoretical and Clinical Aspects of Generalized Anxiety Disorder

Individuals with generalized anxiety disorder are in a constant state of cognitive and behavioral avoidance due to their anxiety or worry. According to Beck, "the anxiety patient, as we have said, avoids only those specific tasks that endanger his vital interests and present some probability of confrontation or failure or of not being able to cope" (2005, p. 104). These avoidances become so habitual or automatic the individual is often not aware or conscious that they are doing this (Orsillo &Roemer, 2011, p. 44). Therefore, it is important to note an individual diagnosed with GAD is so consumed with the cognitions of anxiety and worry that it affects their awareness and influences other perceptions and behaviors.

As stated in Orsillo & Roemer (2011), generalized anxiety disorder leads to a troubled understanding of emotion and ineffective use of emotion, and in turn, creates a negative internalized response to emotions. This internal cognitive response generates a desire to avoid these types of thoughts, putting even more attention and energy onto the exact emotions or thoughts the individual is trying to avoid. "These critical, judgmental reactions to natural responses like fear or doubt amplify and prolong stress" (Haeys-Skelton et al., 2013, p. 264). The constant effort, attention, and resulting self-criticism given to the anxiety and worry become cyclical and make every day life much more difficult to function in and be present for.

Another key feature found within generalized anxiety disorder is expectation of failure. Individuals with generalized anxiety disorder can internalize a situation such as a job interview, public speaking or a social event, where an increase in anxiety is a normal response, as a fearful and threatening event. "In these situations, our biological predisposition to avoid danger and seek safety may not be the optimal responses" (Roemer & Orsillo, 2013, p. 167). According to Kabat-Zinn (2013), if an individual has chronic anxiety, as found in gen-generalized anxiety disorder, the anxiety is often experienced in a manner out of proportion to the actual activity or event. Therefore, an individual can make the ill perceived connection that failure is automatic in these situations and is then linked to avoiding situations in which there is an expected experience of failure. For a person with generalized anxiety disorder, "these life situations pose a threat because 'inadequate' performance makes [an individual] feel constantly vulnerable to negative evaluation and rejection" (Beck et al., 2005, 101).

Accompanying the cognitive components of excessive anxiety and worry, there are also physical and somatic symptoms associated with generalized anxiety disorder. The physical symptoms include muscle tension, feeling shaky, and muscle soreness (APA, 2013). Somatic symptoms connected to GAD can include nausea, diarrhea, sweating, and feelings linked to stress like headaches and irritable bowel syndrome.

It is important to differentiate what the DSM 5 considers generalized anxiety disorder and "nonpathological anxiety" (APA, 2013, p. 222). The DSM 5 states "the worries associated with generalized anxiety disorder are excessive and typically interfere significantly with psychosocial functioning, whereas worries of everyday life are not excessive and are perceived as more manageable and may be put off when more pressing matters arise" (APA, 2013, p. 222). Furthermore, the DSM 5 (2013) argues that worries connected to GAD are inescapable, troubling, and prominent. The DSM 5 claims that nonpathological worries are less frequently associated with physical symptoms.

Theoretical Perspective: Mindfulness

The growing popularity of mindfulness in the United States has further connected mindfulness and psychotherapy, leading to the emergence of using mindfulness practices in conjunction with therapy (Germer, 2013). Mindfulness can include meditation, mindfulness-based stress reduction (MBSR), mindfulness and acceptance-based interventions (MABIs), and formal and informal practices of mindfulness. For the purpose of this paper, mindfulness will be discussed as a perspective or theory in a broader sense and not as a direct therapeutic practice such as meditation or yoga.

Mindfulness Defined

According to Orsillo and Roemer (2011), "the term mindfulness comes from Buddhism, but psychology has begun to recognize that mindfulness (removed from the religious context) may be used to improve physical and emotional well-being" (p. 82). More specifically, Chambers, Gullone, and Allen (2009) state the cognitive aspects of mindfulness stress an awareness of the present experience without judgment. Mindfulness is "an open attention to one's present experience, accompanied by a nonjudgmental, accepting attitude toward whatever one encounters" (Bishop et al, 2004, as cited in Brown, Marquis, & Guiffrida, 2013). In its simplest definition and for further discussion in this paper, mindfulness is "(1) awareness, (2) of present experience, (3) with acceptance (Germer, 2013).

Critique of Generalized Anxiety Disorder Diagnosis from a Mindfulness Perspective

Mindfulness is a theory derived from Buddhism that states having an unbiased awareness of one's experience in the present allows an individual to accept his/her experience, move through the experience, and be less affected by it. This allows the experience to be positive or neutral where it can then be incorporated into the individual's cognition and identity.

On the other hand, according to the DSM 5, a diagnosis of generalized anxiety disorder is based on the underlying assumption that paying attention to or being conscious of one's experience of anxiety, worry, and its associated physical symptoms is distressing to the point of impairment and therefore, pathological. The DSM 5 lists a set of criteria and symptoms for generalized anxiety disorder, which "suggest that the 'symptoms' are the problem (Eifert & Forsyth, 2005, p. 5). Consequently, this type of focus influences how individuals with GAD interpret their own symptoms when they are introduced, assessed, and diagnosed, often times in therapeutic settings. If the criteria and the list of symptoms are what define the disorder, then the person who embodies these criteria and symptoms is "disordered," viewed in a negative manner, and as pathological.

When examining the DSM 5 from a mindfulness perspective, it is apparent the very definition of GAD produces a compounded scenario where "the function of anxiety is to narrow the focus of attention onto threat-relevant information so that only anxiety consistent information is attended to, contributing to a feeling of being defined by anxiety" (Hayes-Skelton et al., 2013, p. 265). Therefore, the DSM 5 criteria narrow the mindset of a person, whose anxiety and worry will only increase, and in turn, worsen his/her anxiety and worry. Mindfulness suggests having an awareness of one's anxiety or worry as it is experienced in the present, but then advocates to let it go from your present mindset and not attach judgment to the anxiety or worry. Anxiety and worry are part of human cognition that mindfulness suggests to experience as they are in the moment. Individuals with GAD attach judgment and perceived threat to everyday anxiety and worry at an extreme and frequency that exacerbates the issue and impairs functioning, according to the DSM 5 (2013). The mere fact the diagnosis is titled "generalized anxiety disorder" places the emphasis on the pathological and the negative by using the word "disorder" to label individuals experiencing these set criteria.

Awareness

Within mindfulness, awareness refers to paying attention and experiencing things as they are (Kabat-Zinn, 2013). It is often times the habit of human nature to be distracted by emotions, activities, and thoughts, and therefore, does not allow one to pay attention to experiences as they are happening, as they are. Individuals regularly attach evaluations and interpretations onto thoughts and emotions, which further detract from awareness. Being mindful "is often challenging for clients at first, because intentionally working with awareness in this way is in direct conflict with habitual modes of human functioning and with many cultural norms" (Brown et al., 2013, p. 98). Contrary to this type of awareness, the DSM 5 diagnosis of generalized anxiety disorder stigmatizes an individual's capacity for paying attention to anything other than this list of criteria viewed as pathological and distressing. Portraying generalized anxiety disorder as a perceived list of criteria and symptoms to be relieved or cured of negates an individual's ability to work beyond anxiety and worry to achieve such awareness found through mindfulness. Based on the criteria given for this diagnosis, the DSM 5 assumes an individual with GAD is merely a sum of his or her dysfunctional parts, making awareness virtually impossible. The DSM 5 claims a person with generalized anxiety disorder is consumed with controlling worry and keeping worry from taking over his/her attention and therefore, having an awareness of one's self, surroundings, and the present experience is not even fathomable according to the defining criteria of this diagnosis.

Being In The Present

When viewing GAD from the mindfulness perspective, there is an immediate disconnect for individuals with this anxiety disorder from experiencing life in the present. Individuals with generalized anxiety disorder are constantly living in the past, connecting past failures and fears to present anxiety, and worrying about the future. The DSM 5 (2013) defines worry as "apprehensive expectation." This can translate into anxious anticipation for something that may or may not happen in some future time. Mindfulness challenges this idea at its core by suggesting that one should pay attention to the present experience (feelings and emotions included) and not to place any cognitive judgment, past or future, onto one's present experience. According to Eifert and Forsyth (2005), "clients with anxiety problems spend an enormous amount of time engrossed in this activity, caught up in an endless pursuit of calm and peace and an ongoing flight from pain and unpleasantness...meanwhile, the world of real experience flows by untouched and untasted" (p. 75). According to the DSM 5, having such distressing anxiety and worry can become all encompassing and make it very difficult for a person with GAD to be in the present. By the very criteria the DSM 5 has pigeonholed clients with GAD with, mindfulness, according to Hayes-Skelton, et al. (2013), brings to the forefront that "the intense cognitive effort associated with monitoring and managing one's internal experience can make it difficult to notice and appreciate the present moment" (p. 265). The criteria presented by the DSM 5 stigmatize individuals with GAD as being incapable of paying attention to or being in the present at all.

Non-judgmental Evaluation

Aspects of mindfulness promote "recognizing that one's thoughts, feelings, and urges are transient internal events rather than inherent, permanent aspects of the self or accurate representations of reality" (Hoge et al., 2015, p. 229). Simply put, thoughts are merely thoughts. The DSM 5 focuses entirely on the anxiety and worry experienced as being all-encompassing, distressing, and causing impairment in functioning for an individual diagnosed with generalized anxiety disorder. This focus could be harmful for individuals or a clinicians treating such a client to take at face value. This could create a situation where anxiety would "be paradoxically increased by experiential avoidance or the internal strategies aimed at suppressing anxious thoughts, feelings, or sensations or changing their form" (Hayes-Skelton et al., 2013, p. 265). Furthermore, "non-judgmental awareness may facilitate a healthy engagement with emotions, allowing individuals to genuinely experience and express their emotions without underengagement (e.g. experiential avoidance and thought suppression) or overengagement (worry and rumination)" (Chambers et al., 2009, p. 566). The DSM 5 systemically categorizes generalized anxiety disorder's defining criteria and symptoms as giving judgmental evaluation to their thoughts and emotions creating anxiety and worry which is then viewed as pathological. On the other hand, mindfulness reminds us these are aspects of human nature and cognition that can be adapted and changed through awareness, kindness, and acceptance.

Conclusion

Generalized anxiety disorder, as defined by the DSM 5, through a list of criteria and symptoms, marginalizes the very people it is supposed to help through diagnosis. While the DSM 5 is widely used for official diagnoses, especially for billing purposes and the practice of doing so, using these criteria only is far from mindful. Examining the diagnosis of GAD from a mindfulness perspective draws attention to how harmful diagnosing only using the DSM 5 can be without proper individual assessment, further research, field experience, cultural sensitivity, and awareness of other perspectives. Mindfulness focuses on a neutral experience to awareness, being in the present moment, and doing so without judgment. The DSM 5 places stigma on individuals diagnosed with generalized anxiety disorder, making the assumption that they are not normal because they experience every day anxiety in extreme and pronounced ways. While the DSM 5 points out pathological indicators for GAD, mindfulness brings to light that the diagnosis should be just that... a diagnosis. The individual who embodies and experiences these specific criteria for generalized anxiety disorder can experience the diagnosis in the present, with an awareness accepting it for what it is, solely as a list of defining criteria to help diagnose, without adding judgment to the individual or by the individual.

References

American Psychiatric Association. (2013). *Diagnostic and statistical manual of mental disorders: DSM 5* (5th ed.). Washington, D.C: American Psychiatric Association

Beck, A. T., Emery, G,. & Greenberg, R. (2005). *Anxiety disorders and phobias: A cognitive perspective.* New York, NY: Basic Books

Brown, A. P., Marquis, A., & Guiffrida, D. A. (2013). Mindfulness-based interventions in counseling. *Journal of Counseling and Development, 91,* 96-104.

Chambers, R., Gullone, E., & Allen, N. B. (2009). Mindful emotion regulation: An integrative review. *Clinical Psychology Review, 29,* 560-572.

Eifert, G. H. & Forsyth, J. P. (2005). *Acceptance and commitment therapy for anxiety disorders: A practitioner's treatment guide to using mindfulness, acceptance, and values-based behavior change strategies.* Oakland, CA: New Harbinger Publications, Inc.

Germer, C. K. (2013). Mindfulness: What is it? What does it matter? In Germer, C. K., Siegel, R. D., & Fulton, P. R. (Eds). *Mindfulness and psychotherapy (2nd ed.)* (pp. 3-35). New York, NY: The Guilford Press.

Hayes-Skelton, S. A., Orsillo, S. M., & Roemer, L. (2013). An acceptance-based behavioral therapy for individuals with generalized anxiety disorder. *Cognitive and Behavioral Practice, 20,* 264-281.

Hoge, E. A., Bui, E., Goetter, E., Robinaugh, D. J., Ojserkis, R. A., Fresco, D. M., & Simon, N. M. (2015). Change in decentering mediates improvement in anxiety in mindfulness-based stress reduction for generalized anxiety disorder. *Cognitive Therapy and Research, 39 (2),* 228-235.

A Comparison of Attachment Theory Versus The Diagnosis of BPD

Lauren Furiasse

Any clinician no matter what setting he or she is in, comes into contact with the *Diagnostic and Statistical Manual of Mental Disorders*, or DSM 5. It is important to retain this material while working with any client to be an efficient professional. However, there are many other factors that need to be considered for an individual receiving a diagnosis. That is why it is important to be educated about other theories with a more detailed explanation about the client's diagnosis than the DSM 5.

One type of theory focused on in this paper is attachment theory. Attachment theory has been developed and utilized by many in multiple settings for many years. The most widely known researchers associated with the development of attachment theory are John Bowlby and Mary Ainsworth. The goal of this paper is to provide an understanding for the reader about how Borderline Personality Disorder can be a natural response for an adult who has an insecure attachment style.

Diagnostic Criteria

Before discussing the role attachment theory plays for an individual with Borderline Personality Disorder, a breakdown of this diagnosis would create a better perception of what this diagnosis looks like for an individual. According to the American Psychiatric Association in Diagnostic And Statistical Manual Of Mental Disorders: DSM 5 (2013), Borderline Personality Disorder or BPD must possess certain features to be classified as BPD over any other disorder. These features are a pervasive pattern of instability of interpersonal relationships, self-image, and affects, and marked impulsivity (American Psychiatric Association, 2013). These patterns must begin by early adulthood and be present in a variety of contexts (American Psychiatric Association, 2013).

In order for an individual to meet these criteria, someone must possess five or more diagnostic criteria identified as Borderline Personality Disorder (American Psychiatric Association, 2013). The first criterion is someone with

BPD symptoms would create frantic efforts to avoid real or imagined aban-donment (American Psychiatric Association, 2013). However, under this section it does not include suicidal or self-mutilating behavior (American Psychiatric Association, 2013). Individuals with this diagnosis contain a perception of impending separation or rejection, or loss of external structure (American Psychiatric Association, 2013). This perception can cause the individual to have profound changes in self-image, affect, cognition, and behavior (American Psychiatric Association, 2013). Not only does someone with BPD experience this perception, he or she is sensitive to environmental circumstances (American Psychiatric Association, 2013).

As a clinician moves forward with the diagnostic criteria, the second criterion must deal with interpersonal relationships. Someone with BPD diagnosis can possess a pattern of unstable and intense interpersonal relationships (American Psychiatric Association, 2013). These interpersonal relationships are characterized by alternating between extremes of idealization and devaluation (American Psychiatric Association, 2013). Someone can also have identity disturbance demonstrated as a self-image or sense of self that is unstable and, this instability of self would be persistent (American Psychiatric Association, 2013).

The next area of the criteria an individual with BPD can display is having impulsivity in two areas (American Psychiatric Association, 2013). This impulsivity can lead to be self- damaging to this individual (American Psychiatric Association, 2013). The impulsivity can be seen as spending money, sex, substance abuse, reckless driving, and binge eating (American Psychiatric Association, 2013). However, suicidal or self-mutilating behavior are not defined as self-damage in this section (American Psychiatric Association, 2013). The suicidal or self-mutilating behavior, gestures, or threats are their own criteria due to recurrence (American Psychiatric Association, 2013).

The last section for the criteria for BPD is affective instability due to marked reactivity in mood, chronic feelings of emptiness, inappropriate anger, and stress-related paranoid ideation (American Psychiatric Association, 2013). The instability for this individual is caused by reactivity of mood (American Psychiatric Association, 2013). This would be seen as intense dissatisfaction, irritability, or anxiety that only lasts for a few hours (American Psychiatric Association, 2013).

One key feature of interest displayed in someone with BPD is anger. The anger shown in this diagnosis would be displayed as temper, constant anger, recurrent physical fights (American Psychiatric Association, 2013). This anger can be shown at times when faced with realistic time-limited separation or when there are unavoidable changes in plans (American Psychiatric Association, 2013). An unavoidable change in plan can be when a clinician announces the end of the hour (American Psychiatric Association, 2013). An individual with BPD would respond to this unavoidable change with despair (American Psychiatric Association, 2013). Anger is not the only response a client would display with unavoidable change. Panic or fury can be an immediate response when someone important to them is just a few minutes late or must cancel an appointment (American Psychiatric Association, 2013). The main theme in this section would be the inappropriate behavior or difficulty managing his or her anger in response to something inevitable happening in his or her life.

All of these components create the criteria for the diagnosis of Borderline Personality Disorder. There is a certain theme appearing within this diagnosis leading to other factors within this diagnosis. This theme is the idea of abandonment for these individuals. A client who meets the first criteria has "Frantic efforts to avoid real or imagined abandonment" (American Psychiatric Association, 2013). They must believe that this "abandonment" implies they are "bad" (American Psychiatric Association, 2013). This type of fear surrounded by the idea of "abandonment" is related to an intolerance of being alone and a need to have other people with them (American Psychiatric Association, 2013).

This individual would go through extreme efforts to prevent this abandonment (American Psychiatric Association, 2013). These efforts can be self-mutilating or suicidal behaviors. (American Psychiatric Association, 2013). The criteria for BPD correlates with the interaction of others in that individual's life, whether this interaction be through relationships, communication, or emotional responses towards others. The key factor is the reaction or response this individual has to the relationships in his or her life. However, in order to provide a proper answer about why this individual is diagnosed with BPD, a look into a person's attachment style will be of great importance to see if there are any similarities between this person's diagnosis and attachment style.

Attachment theory

Any professional reading the DSM would be able to see a commonality emerging in people who are given the diagnosis of BPD. According to Innerhofer (2013), forty seven percent of all patients with a fearful preoccupied attachment style were diagnosed with BPD. This seems to be a high statistic in regards to one type of attachment style appearing in patients with PBD. In order to grasp what this means for an individual with this type of attachment, one must retain a better insight into the breakdown of this theory.

According to Joan Stevenson-Hinde in "Attachment Theory And John Bowlby: Some Reflections" (2007), Bowlby stated that attachment theory is "The inner world of emotion understanding, and the capacity to speak freely and openly about negative and positive feelings." These emotions are uniquely related to the mother-child relationship (Hinde, 2007). Bowlby proposed this early relationship with a primary caregiver, usually the mother, leads to "Generalized expectations about the self, others, and the world" (Waters, Hamilton, & Weinfield, 2000). The representations of these expectations cognitively are known as "working models" (Waters et al., 2000). These representations will emerge early in development and will evolve during the attachment-related experiences during childhood and adolescence (Waters et al., 2000).

As this relationship between the mother and her child begins to form, this interaction can affect many other parts of the child's development. Bowlby believed the one area of development that can be impacted was emotional communication (Hinde, 2007). The process for the self, others, and the motivational state of the individual is all shaped by this emotional communication (Hinde, 2007).

During the beginning years of development, the pattern of communication a child adopts towards his or her mother matches the pattern of communication that the mother has been adopting towards his or her child (Hinde, 2007). These two individuals begin a mirroring process. The importance of communication a child establishes is at the core of the development and maintenance of attachment. (Hinde, 2007). Sophia Landa Hons and Robbie Duschinsky in "Letters From Ainsworth: Contesting The Organization Of Attachment" (2012), describes how Bowlby observed this disruption and separation from an attachment figure. This separation becomes the source of anxiety for the child (Hons & Duschinsky, 2012). This anxiety can trigger the type of attachment system the child has with his or her caregiver (Hons & Duschinsky, 2012). The takeaway from John Bowlby's observations for attachment was how vital the mother or any caregiver is to the child.

The first relationship anyone develops is during his or her childhood. Whoever the child identifies as his or her caregiver will determine his or her attachment style. How the child and caregiver interact with each other will set the way for how that child interacts and establishes relationships for the rest of that child's life. Holmes describes Bowlby supporting the notion of individuals maintaining the need for a secure base once people enter into adulthood. Bowlby viewed this development in terms of a move from immature to mature dependence (Holmes, 2004). At times of extreme stress, people turn to someone who is seen as his or her secure base (Holmes, 2004). The need for a secure base does not change once anyone reaches adulthood.

Once John Bowlby embarked on the idea of attachment, other researchers began to expand this theory. Mary Ainsworth began her own observation that separation from an attachment figure would affect the entire attachment system (Hons & Duschinsky, 2012). Ainsworth created a structured situation to observe fifty-six middle class infants all around the age of eleven months (Hons & Duschinsky, 2012).

Ainsworth's goals were to examine individual differences these infants possessed when they responded to the departure of a caregiver (Hons & Duschinsky 2012). This experiment came to be known as "The Strange Situation Procedure" (Hons & Duschinsky, 2012). Ainsworth was able to expand Bowlby's idea of attachment by creating names and classifications for these different attachment styles. In order to achieve this classification, Ainsworth observed children interacting with their caregivers. Infants were classified as Anxious- Avoidant, (Group A), Secure (Group B), Anxious-and Ambivalent/ Resistant (Group C) (Hons & Dushinksky, 2012).

Infants who showed no distress on separation, and ignored the caregiver once the caregiver returned were classified as Anxious-Avoidant (Group A) (Hons & Duschinsky, 2012). Infants who were Secure (Group B) used the caregiver in a way as a safe base to explore (Hons & Duschinsky, 2012). These young children protested the departure from his or her caregiver, but would seek for his or her caregiver once the caregiver would return to them (Hons & Duschinsky, 2012). Infants classified as Ambivalent/Resistant (Group C) showed distress on separation and were hard to comfort on the caregiver's return (Hons & Duschinsky, 2012). Ainsworth's Strange Situation Procedure was found to be applicable to a large majority of infants from middle-class families and can be applied cross- culturally (Hons & Duschinisky, 2012).

Ainsworth noted from this experiment that abused-and-neglected infants show a combination of the A/C pattern (Hons & Duschinsky, 2012). However, Ainsworth had a small group of infants known as "Unclassified" (Holmes, 2004). These infants were subject to their caregiver's unresolved loss or trauma (Holmes, 2004). The caregiver is triggered by painful memories of her own childhood (Holmes, 2004). These memories prevent the caregivers ability to maintain affective continuity in creating a secure base for the child (Holmes, 2004).

Birgit Innerhofer (2013), states Ainsworth's experiment observed that a bond was activated between the caregiver and his or her child. This bond was during a distressing situation such as separation from their caregiver, or when anxiety, fear, distress or sadness arose (Innerhofer, 2013). Ainsworth was able to depict a pattern and the effects different relationship interactions caregivers had on their infants. From this observation, Ainsworth was able to provide answers for future researchers using attachment theory through this classification system.

Attachment Theory or DSM 5

After reviewing the DSM and literature for attachment theory, both of these features are trying to seek an answer for certain psychological responses people display with a BPD diagnosis. Whether this behavior is diagnosed as BPD or an attachment problem, the importance is which one provides a better solution for the individual displaying this type of behavior. After working with individuals from all ages, ethnicities, and class systems in social services, the overall impression the diagnosis of BPD creates for an individual is a label rather than a solution.

Throughout my experience in multiple social service agencies, the diagnosis of BPD creates more of a stigmatization for the individual. The diagnosis of BPD does more harm than good for this person. It would be more beneficial for the individual seeking help to be seen as a person rather than to be identified as his or her BPD diagnosis. In order for this to happen, the way professionals view someone's psychological disorder needs to be seen in regards to a psychosocial approach. Instead of just focusing on an individual's physical reactions and responses as a problem, it should be seen as answers into the understanding of that individual. A way of achieving this is by utilizing attachment theory in the diagnosis of BPD

While going over the diagnosis of BPD again, having an attachment theory outlook changes the look of this diagnosis. The first criterion states someone must display a pervasive pattern of instability of interpersonal relationships, self-image, and affects, and marked impulsivity (American Psychiatric Association, 2013). This pattern would be named as an insecure attachment style emerging in these patients. Even though the DSM 5 is trying to name this behavior, there still lacks an answer for why this person displays this certain pattern of behavior. In order to achieve this explanation the utilization of the person's early attachment style would be able to provide a better explanation for the behavior.

Information regarding attachment stresses the importance of the formation of a base whether secure or not in childhood. It would make sense for this persistent pattern of instability of interpersonal relationships, self-image, and affect to begin to show by early adulthood in the diagnosis of BPD in a variety of contexts (American Psychiatric Association, 2013). Since the need for a secure base never vanishes, these individuals are constantly seeking someone to identify as his or her secure base in multiple settings (Holmes, 2004). These settings can be at work, friendships, or personal relationships. No matter the setting for an individual with an insecure attachment. They can display this persistent pattern in order to create a secure base. A therapist working with this adult would know this individual has an insecure attachment style. This answer would provide more education for a therapist working with this client. The therapist can become the secure base for the client eliminating this ineffective persistent pattern this person is displaying in other settings.

While working in a mental health center, my observations were that many clinicians would use the client's diagnosis as an explanation for any of the client's problems. Instead, clinicians can use this information as a benefit to create a stronger relationship with clients if clinicians would discuss the client's early attachment style. Children and adults with an insecure attachment style are more likely to show evidence of psychological disorders than those with a secure attachment style (Innerhofer, 2013). This information will be helpful in identifying that the majority of individuals being diagnosed with BPD would display an insecure attachment style. Individuals with a secure attachment style did not suffer from instability from their secure base preventing them from having difficulties with future emotions, communications, and interpersonal relationships into adulthood.

This type of interaction will be emulated into other relationships leading into this child's adulthood. This attachment style will be emulated through the client's attachment style he or she had with his or her primary caregiver. If a child does not experience positive validation or emotional attendance from his or her primary caregivers in the forms of hugs, cuddles or praise, that child's emotional needs remain unmet and he or she is more likely to internalize emotional distance (Innerhofer, 2013). As someone reaches early adulthood, these types characteristics have already been formed due to the early relationship this individual had with the primary caregiver. Instead of automatically going to the DSM for an answer, clinicians should utilize the attachment theory as an explanation for client's behaviors in regards to BPD.

In addition to the finding noted above that forty seven percent of all patients with a fearful preoccupied attachment style were BPD patients, this attachment style is also often linked to emotional or dysthymic problems and anxiety (Innerhofer, 2013). These symptoms are also part of the criteria for the diagnosis. The DSM states "Affective instability due to marked reactivity of mood such as intense episodic dysphoria, irritability, or anxiety" (American Psychiatric Association, 2013).

Since the need for a secure base never leaves someone, looking at the first criterion someone with BPD symptoms would demonstrate is this individual can create frantic efforts to avoid real or imagined abandonment (American Psychiatric Association, 2013). These individuals as children lacked a care-giver who could validate their internal world and see them as autonomous and sentient all of which produce temporary physiological features associated with secure base experience (Holmes, 2004).

According to the DSM, impulsivity in at least two areas such as binge eating, substance abuse and recurrent suicidal behavior can be shown in order to be diagnosed (American Psychiatric Association, 2013). Someone with an unresolved/preoccupied attachment style would display psychobiological responses to threats of separation or disconnection by self-injury, bingeing and vomiting, and substance abuse (Holmes, 2004). Therefore this individual engaging in this type of behavior would be identified as having unresolved/preoccupied attachment style. This type of individual does not have a secure base, and threats of separation or disconnection would cause displays of psychobiological responses to threats of separation or disconnection by self-injury, bingeing and vomiting, and substance abuse (Holmes, 2004).

Therefore, this individual engaging in this type of behavior would be identified as having unresolved/preoccupied attachment style. This type of individual does not have a secure base and threats of separation or disconnection are leading to this individual to respond in a self-inflecting behavior. The way to prevent this behavior is to provide a secure base so this individual does not experience feelings of separation or disconnection. The DSM 5 would not be able to provide reasons for the reactions of this person.

Overall, difficulties people face can have multiple solutions. It is vital to someone's recovery to have insight into the person's biology, environment, physical, and psychological instead of just placing a diagnosis onto this person. A way of achieving an understanding is by utilizing attachment theory when someone is displaying BPD symptoms. This theory would provide a more in-depth understanding of this person's difficulties to be able to effectively help them.

References

American Psychiatric Association. (2013). Diagnostic and statistical manual of mental disorders: DSM 5. Washington, D.C: American Psychiatric Association.

Holmes, J. (2004). Disorganized attachment and Borderline Personality Disorder: A clinical perspective. *Attachment & Human Development, 6*(2), 181-190.

Innerhofer, B. (2013). The relationship between children's outcomes in counseling and psychotherapy and attachment styles. *Counseling Psychology Review, 28*(4), 60-76

Landa, S., & Duschinsky, R. (2013). Letters from Ainsworth: Contesting the 'Organization' of Attachment. *Journal Of The Canadian Academy Of Child & Adolescent Psychiatry, 22*(2),172-177.

Stevenson-Hinde, J. (2007). Attachment theory and John Bowlby: Some reflections. *Attachment & Human Development, 9*(4), 337-342

Waters, E., & Hamilton, C. E. (2000). The Stability of Attachment Security from Infancy to Adolescence and Early Adulthood: General Introduction. *Child Development, 71*(3), 678.

Borderline Personality Disorder and Attachment Theory

Brigid Geary

Borderline Personality Disorder (BPD) has been perceived as a circumstance of symptoms related to those who have unstable and chaotic relationships, trouble with emotional regulation, and difficulty functioning on their own (Steele, Bate, Nikitiades, & Buhl-Nielsen, 2015). The development of Borderline Personality Disorder is linked to early childhood experiences, relationships with caregivers, and traumatic life events (Steele et al., 2015). This paper will focus on Borderline Personality Disorder as a consequence of disorganized attachment in childhood with a caregiver. It will also include a clinical view on how this understanding of the disorder may impact a client when making a diagnosis. It will begin with a description of Borderline Personality Disorder according to the *Diagnostic and Statistical Manual of Mental Disorders* (American Psychiatric Association, 2013), followed by a description of attachment theory and the link between the Borderline Personality Disorder criteria and attachment theory. Finally, this paper will end with a critique of the use of attachment theory in making a diagnosis of Borderline Personality Disorder.

Borderline Personality Criteria

According to the DSM- 5, Borderline Personality Disorder is associated with a persistent pattern of unstable interpersonal relationships, self-image, and affects, as well as a marked impulsivity, which present in a variety of contexts and with onset in early adulthood (American Psychiatric Association, 2013). There are nine criteria in the DSM 5 for Borderline Personality Disorder, however, only five of the criteria need to be met to make a diagnosis (American Psychiatric Association, 2013). According to the DSM 5 (American Psychiatric Association, 2013), those with Borderline Personality Disorder make frantic efforts to avoid real or imagined abandonment (Criterion 1). For example, if a person with Borderline Personality Disorder is waiting for someone to pick them up, and the individual is late, they believe that they are being abandoned and label the other as a "bad" person (American Psychiatric Association, 2013).

According to the DSM 5 (American Psychiatric Association, 2013), BPD individuals have a pattern of unstable and intense relationships (Criterion 2). They may demonstrate a high level of self- disclosure to caregivers or lovers when first meeting them and demand to spend much time together, however, they may quickly switch to believing that the other individual does not care about them. According to the DSM 5 (American Psychiatric Association, 2013), BPD individuals may have an identity disturbance characterized by a marked and persistent unstable self- image or sense of self (Criterion 3). They may demonstrate sudden shifts in their outlook on goals, values, or careers. When BPD individuals believe that they are not receiving enough support in a relationship, they may have feelings that they do not even exist. For example, they may feel nothing because they do not believe they have any support or have any feelings in a relationship. In turn, their performance in work and school settings may be jeopardized due to their sudden thought change.

Moreover, individuals with BPD display impulsivity in at least two areas that may be damaging (Criterion 4) (American Psychiatric Association, 2013). Such impulsivity may present as behaviors such as gambling, binge eating, engaging in unsafe sex, or spending money excessively. BPD individuals frequently display recurrent suicidal behaviors, gestures, or threats, or engage in self-mutilating behaviors (Criterion 5) (American Psychiatric Association, 2013). Often, these behaviors are attempts to avoid imagined or real abandonment. Suicide attempts and threats are very common, with successful suicides occurring in 8%-10% of BPD individuals (American Psychiatric Association, 2013). Additionally, individuals with BPD may display affective instability that is due to a marked reactivity of mood (Criterion 6; American Psychiatric Association, 2013). If present, this period of affective instability is brief, typically lasting several hours but not usually for longer than a few days. Such emotional lability may be understood as stemming from their stressful interpersonal experiences.

Individuals with Borderline Personality Disorder may be troubled by chronic feelings of emptiness (Criterion 7) (American Psychiatric Association, 2013). Individuals diagnosed with BPD often have low thresholds for boredom, which may present as an active pursuit for new and stimulating experiences. According to the DSM 5 (American Psychiatric Association, 2013), they frequently express inappropriate and intense anger or have difficulty controlling this anger (Criterion 8). It is when BPD individuals perceive caregivers or lovers as neglecting or abandoning them that this anger often arises. Additionally, during times of extreme stress, transient paranoid ideation or dissociative symptoms may occur (Criterion 9; American Psychiatric Association, 2013). Similar to the symptomology of the other diagnostic criteria, paranoid ideation or dissociation frequently occurs as a consequence of the experience of perceived or threatened abandonment (American Psychiatric Association, 2013).

This aforementioned description of the diagnostic criteria for Borderline Personality Disorder offers a brief outline of the clinical symptoms that will be useful in the context of understanding attachment theory, which is described in the subsequent section. More specifically, attachment theory is central to understanding the development of BPD symptoms. Although insecure attachment may play an active role in the development of Borderline Personality Disorder, of equal importance are the ways that insecure attachment is experienced.

Description of Attachment Theory

Attachment theory was created by John Bowlby, who found that in times of distress individuals typically need physical or emotional attachment to others, which provides a sense of security and feelings of safety (Steele, Bate, Nikitiades, & Buhl-Nielsen, 2015). The attachment figure is argued to be a secure base for the child; the child is able to temporarily leave the caregiver knowing that the caregiver will warmly receive the child when he or she returns, which is especially important for a child in a frightful situation (Steele et al., 2015). John Bowlby (as cited in Steele et al., 2015) suggests recognizable elements of attachment behavior that keep children close to their caregivers. The relational components that maintain a strong connection between a child and a caregiver include smiling or crying, the child maintaining physical proximity to the caregiver, and the subsequent formation of internal working models by age three (Fonagy, Target, Gergely, Allen, & Bateman, 2003).

A child uses four representational systems to predict the behaviors of others or the self's behavior simultaneously with more natural states from various situations the child experiences (Fonagy et al., 2003). These systems include the connection that is made between the mother and the child in the first year of a child's life; the object representations of the child's attachment; autobiographical memories; and evaluations of others' mindsets that the child uses to make comparisons to thoughts of the self (Fonagy et al., 2003).

Attachment theory was further developed following a study conducted by Mary Ainsworth that observed attachment patterns in infants who were 12-18 months old (Steel et al., 2015). The strange situation study found that different patterns of attachment exist, namely, secure, insecure-avoidant, and insecure-resistant (Steele et al., 2015). A secure attachment meant that it was clear to the child that his or her parent is available to him or her when he or she is in distress (Steele et al., 2015). Additionally, insecure-avoidant attachment was demonstrated when the child turns to the parent when in distress and is rejected but the child finds other coping strategies instead of turning to the parent (Steele et al., 2015). Finally, insecure-resistant attachment was demonstrated when the child remained in distress even when connected back with the parent (Steele et al., 2015). The strange theory paradigm underscores the importance of a long- term secure base.

In an examination of the strange situation paradigm, patterns of attachment presented that did not fit into one of the three aforementioned categories, thereby leading to the development of a fourth attachment pattern, disorganized attachment (Steele et al., 2015). Disorganized attachment occurred when the caregiver inconsistently responded to the needs of a distressed child (Steele et al., 2015). Children with disorganized attachment, in times of distress, may separate from the parent or fail to present with emotional reactions. This attachment pattern is termed disorganized because the child is in fear of the caregiver, but at the same time they are the only individual they can turn to (Steele et al., 2015). Moreover, this type of attachment is most commonly observed in abusive households (Steele et al., 2015). Before diving into my clinical perspective of this link, I will discuss a study done that connects early childhood attachments and Borderline Personality Disorder.

Link between BPD and Attachment Theory

The link between Borderline Personality Disorder and attachment theory has been studied by connecting early attachment to BPD developing in adolescence and adulthood. Carlson et al. (2009) theorized that the parent-child attachment shapes the development of a child's experiences and feelings. It has been found that disorganized and insecure parent-child attachments are associated with inadequate communication and emotional regulation during childhood (Steele et al., 2015).

Carlson and colleagues (2009) studied a group of 162 individuals from birth to age 26. The researchers were seeking to establish and specify the biological and environmental factors early in life that influence an individual's sense of self, where the self is a mediator between these different factors and Borderline Personality Disorder (Carson et al., 2009). In other words, the disorder is detached from biological and environmental factors when the sense of self is not taken into consideration. For biological factors, they evaluated medical history at birth, behavior and temperament (Steele et al., 2015). For environmental factors, they evaluated the mother's relationship status and possible maltreatment (Steele et al., 2015).

The individuals were assessed at five different periods. When the individuals were 12-18 months, they observed attachment using the strange situation paradigm and noted the mothers' behaviors (Steele et al., 2015). In middle school, they assessed children's views regarding their families through drawings. At the age at 12, the children participated in a narrative assessment. (Steele et al., 2015). At the age of 13, they interacted with family by structured activities (Steele et al., 2015). Lastly, at age 26, a diagnostic interview, self-report measures of symptoms, and health were all assessed (Steele et al., 2015). Throughout this study, family disruptions and stressful life events were noted (Steele et al., 2015). The results showed a relationship between individuals who have BPD symptoms and family and child variables in childhood, such as disorganized attachment and maltreatment, maternal hostility, family issues, and family life stress (Steele et al., 2015).

These results indicate that individuals with disorganized attachment at an early age are more likely to present with symptoms of BPD (Steele et al., 2015). I believe that this makes sense, such that an individual who was maltreated as a child becomes someone who cannot keep stable relationships because they grew up in families plagued with violence and mistrust. A person with Borderline Personality Disorder may desire a stable relationship with a lover or caregiver, however, his or her past experiences influence him or her in ways that limit relational capacities.

Fonagy, Target, Gergely, Allen, and Bateman (2003) explored Borderline Personality Disorder patients' histories of attachment. According to the individuals with a history that involved threats of abandonment, many of them demonstrated a preoccupied attachment (Fonagy et al., 2003). A preoccupied attachment is when the child shows anxiety and anger around their caregiver or lover (Fonagy et al., 2003). This type of attachment can be closely related to a disorganized attachment because both of them are rooted in forms of mal-treatment or trauma that occurred in their lives from a lover or caregiver.

Critiquing Attachment Theory and BPD

In analyzing the link between Borderline Personality Disorder and attachment theory, it is important to evaluate this relationship with a clinical lens. Such is accomplished through a critique of attachment theory as it related to BPD. A narrow focus on a disorganized attachment may lead to an insufficient report on a client's case. The subsequent section of this paper presents four reasons that a large emphasis on attachment theory may lead to an over diagnosis of BPD.

Firstly, anxious attachments are common. Milyavskaya & Lydon (2013) found in a study on attachment that over a quarter of their participants were characterized as having an insecure attachment with their caregivers. An individual who demonstrates insecure attachment may show signs of it in adulthood by being very on edge about keeping a relationship. Therefore, when a client comes to a clinician with signs of both an insecure attachment and symptoms of Borderline Personality Disorder, it is essential to explore the client's life and how this insecure attachment has affected them. For example, a client may be viewed as someone who had an insecure relationship with his or her parents in childhood, however, he or she may have had a specific experience with this insecure attachment due to the environment. The specific experience with this attachment may have been different from others, especially based on variety in developmental factors (Fonagy et al., 2003).

Developmental factors from birth on are important to consider, as they could determine how the client experienced his or her insecure attachment. A critical developmental factor in a child is developing an agency of the self (Fonagy et al., 2013). This includes the child realizing that the way one communicates impacts others, that their physical actions can produce change, that different options exist to accomplish a goal, and that their actions are influenced by prior experiences (Fonagy et al., 2013). An insecure attachment creates barriers in development, such as not being able to reflect on personal feelings in a way that is not overwhelming for the individual (Fonagy et al.,

2003). This lack of reflection can lead an individual to have trouble distinguishing the difference between reality and fantasy, and between behaviors and thoughts. Individuals with BPD may believe that their loved one is abandoning them and in turn, wants to physically hurt them. If thoughts are not differentiated from behaviors, the individuals are more likely to become violent toward their lover. Although this developmental component presents as a risk factor for developing Borderline Personality Disorder, in insecure attachment, the child often finds ways to differentiate between the two in adolescence or adulthood.

Secondly, although there may be an insecure attachment in childhood, individuals with Borderline Personality Disorder may develop different ways to cope with the fear and anxiety around being abandoned (Fonagy et al., 2003). On the other hand, those who have always felt a sense of security may be unaware of how to handle a situation where they are completely abandoned by their caregiver for a long period of time. Therefore, when assessing for BPD, exploring the coping skills an individual has when presented with an abandonment situation is an important component. These coping skills will inform the clinician of the extent of the effect that these symptoms have on their daily functioning. Without looking at these coping skills, it could lead to the individual is being misdiagnosed because of the focus on an insecure attachment in childhood.

For example, a patient who presents with clear signs of childhood maltreatment and exhibits five of the BPD symptoms may automatically be diagnosed with BDP. From a different perspective, observing the client's coping strategies to manage these symptoms may led to a change in diagnosis. If the client makes strong efforts to avoid abandonment, they may have the capacity to utilize other coping mechanisms, such as spending time with relatives, going to the gym, or thinking about the situation in a positive way. Exploring the ways that a client copes with such symptoms could help prevent an over diagnosis of the disorder. Although the patient's symptoms appear to be negatively affecting his or her life, the underlying components of the symptoms are crucial in order to get an accurate diagnosis.

They could also cope with these symptoms by knowing that they are able to manage such fears of abandonments, similar to the strategies used in the past. A client that comes to a clinician with symptoms of Borderline Personality Disorder should be assessed by the client's view of how past experiences currently affect interpersonal relationships. Similarly, client subjectivity is important in distinguishing if these symptoms are truly affecting their relation-relationships. If the clinician only has a handle on the client's past history of unstable relationships, a big part of the picture could be missing.

Thirdly, those who are diagnosed with Borderline Personality Disorder come from all different types of relationships and situations (Fonagy et al., 2003). This suggests that a client who had a secure relationship with his or her caregiver throughout their childhood could potentially develop Borderline Personality Disorder as well. As clients are assessed, Borderline Personality Disorder should not be ruled out because they did not have an insecure attachment in childhood. Their secure attachment in childhood could have been a protective factor; however, another experience with someone other than their caregiver could have been a trigger to the symptoms. For example, they may have had a secure attachment with the caregivers, however, had a traumatic relationship with a teacher in their childhood that affected their outlook on relationships. Clinicians need to be aware of the client's entire situation in order to make the correct diagnosis.

Moreover, as discussed by Fonagy et al. (2003) a patient with a diagnosis of Borderline Personality Disorder may show signs of early insecure relationships, only because they are highly influenced by the emotions of family members and the clinician. If a patient comes in with his or her family and is overwhelmed by the emotions that his or her family is exhibiting, his or her symptom presentations may enhance. The clinician may interpret this as a sign of disorganized attachment. A patient with Borderline Personality Disorder may also be highly affected by the clinician's emotions. Therefore, when assessing or counseling an individual with a potential BPD diagnosis, it is important for the clinician to remain aware of his or her emotions. If these symptoms are enhanced, it may lead the client to show more symptoms that they do not normally show on a daily basis.

Lastly, Borderline Personality Disorder patients typically show violent behavior on either themselves or another individual (Fonagy et al., 2003). A study done by Agrawl, Gunderson, Holmes, and Lyons-Ruth (2004) found that many Borderline Personality Disorder individuals who have an insecure attachment do not have resolution from features that were present in their caregivers. These resolutions may consist of exploring the reasons a client's mother or father was so violent or the reason why they never comforted the client in times of distress. If a client displays such violent behaviors during assessment, it would be beneficial to discuss the characteristics of the client's caregivers to form a resolution for the patient. Clients who were maltreated as children with no resolution of the situations frequently produce the belief that they did something wrong. When giving the diagnosis of Borderline Personality Disorder a clinician needs to be aware of the lack of resolution that may have occurred because this could help the client right from the start.

More importantly, this violent behavior may be caused by environmental factors. The client may have symptoms of depression or anxiety and express it in a violent way. Moreover, it is important to have an open mind as a clinician to all of the different factors that cause an individual to become violent. The client could even be overly stressed and not know how to know to handle hard times in his or her life.

The criteria for Borderline Personality Disorder in the DSM 5 are helpful in order to make a potential diagnosis for a client. Attachment theory should be explored within a patient with Borderline Personality Disorder due to the impact an insecure attachment has on an individual as they move into adulthood. Their inability to keep relationships with lovers and caregivers in adulthood is very understandable as they have never known what a stable relationship has looked like and have lost trust in everyone around them.

Bringing it all together, although an insecure attachment is a precursor to developing Borderline Personality Disorder, clinicians need to be aware of the various factors that go into an insecure attachment. Each client experiences an insecure attachment in a different way because no one's environment is exactly the same. Moreover, a client could have a secure attachment and still develop Borderline Personality Disorder, and they could also have an insecure attachment to a lover or caregiver and have adequate coping skills to deal with it. In conclusion, an insecure attachment in childhood has been shown to have a strong link to the development of Borderline Personality Disorder. Although this has important clinical implications, particularly regarding diagnosis, an examination of the individual with consideration of his or her unique situation is essential in order to establish an accurate diagnosis.

References

Agrawal, H. R., Gunderson, J., Holmes, B. M., & Lyons-Ruth, K. (2004). Attachment studies with borderline patients: A review. *Harvard Review of Psychiatry* (Taylor & Francis Ltd), 12(2), 94-104.

American Psychiatric Association. (2013). *Diagnostic and statistical manual of mental disorders* (5th Ed.). Washington, DC: American Psychiatric Association.

Fonagy, P., Target, M., Gergely, G., Allen, J. G., & Bateman, A. W. (2003). The development roots of borderline personality disorder in early attachment relationships: A theory and some evidence. *Psychoanalytic Inquiry, 23*(3), 412.

Lansky, M. R. (2003). Discussion of Peter Fonagy et al.'s "The Developmental Roots of Borderline Personality Disorder in Early Attachment Relationships: A Theory and Some Evidence". *Psychoanalytic Inquiry, 23*(3), 460

.Milyavskaya, M., & Lydon, J. E. (2013). Strong but insecure: Examining the prevalence and correlates of insecure attachment bonds with attachment figures. *Journal of Social and Personal Relationships, 30* (5), 529-544.

Steele, M., Bate, J., Nikitiades, A., & Buhl-Nielsen, B. (2015). Attachment in adolescence and borderline personality disorder. *Journal of Infant, Child, and Adolescent Psychotherapy*, 14:16-3.

A Cognitive-Behavioral Approach to Anorexia Nervosa in the DSM 5

Mara Maeglin

Anorexia Nervosa in the DSM 5

Criterion A: Low-Body Weight

According to the *Diagnostic and Statistical Manual of Mental Disorders* (DSM 5; American Psychiatric Association 2013), three central features characterize anorexia nervosa (AN), the first of which being a persistent restriction of energy intake that results in a significantly low body weight. Significantly low body weight is defined as, "a weight that is less than minimally normal, or for children and adolescents, less than minimally expected (American Psychiatric Association 2013, p. 338). The DSM 5 acknowledges the difficulties in assessing one's weight, noting the variance among individuals' normal weight ranges and the discrepancies between definitions for thin and underweight thresholds (American Psychiatric Association 2013).

The current severity specification for AN outlines four categories that are correlated with ranges of body mass index (BMI) derived from World Health Organization categories for thinness in adults (American Psychiatric Association 2013) Body Mass Index is calculated as weight in kilograms/height in meters^2. Accordingly, adults with a BMI that is greater than or equal to 18.5 kg/m^2 are not considered to meet the criteria for a significantly low body weight (American Psychiatric Association 2013, p. 340). The World Health Organization (as cited in American Psychiatric Association 2013, p. 340) defines "moderate and severe thinness" by a BMI that is less than17.0 kg/m^2, and the DSM 5 uses this measure to define significantly low body weight. Thus, the symptom severity categories include the following: mild (BMI $\geq$ 17 kg/m^2); moderate (BMI 16 – 16.99 kg/m^2); severe (BMI 15- 15.99 kg/m^2); and extreme (BMI < 15 kg/m^2) (American Psychiatric Association 2013, p. 339).

Given the nature of development, children and adolescents are expected to gain weight as they grow thus eliciting different standards for determining significantly low weight. Due to differences in developmental trajectories, no clear numerical guidelines exist for defining normal weight in children and adolescents. The DSM 5 uses BMI-for-age percentiles for adolescents and children that correlate with the symptom severity BMI ranges for adults. The Center for Disease Control and Prevention (as cited in American Psychiatric Association 2013, p. 340) considers a BMI-for-age that is below the fifth percentile as "underweight." However, the DSM 5 suggests that children who are above this standard may be considered underweight if they do not meet their expected growth trajectories (American Psychiatric Association 2013).

It is important to note that individuals with anorexia nervosa restrict their energy intake in various ways. They often have extreme dietary rules that dictate eating behaviors, such as when to eat (e.g., never before dinner), what to eat (e.g., only fruits and vegetables), and how much to eat (e.g., very small portions; Grave 2013). Additionally, food checking is a common behavior among individuals with anorexia nervosa. These behaviors may occur during or outside meals, and often manifest as counting calories (Grave 2013). Additionally, individuals with AN will frequently dispose of food in napkins or hide food in their pockets during meals, cut food into small pieces, and spend much time rearranging the pieces of food on their plates (Sadock, Sadock, & Ruiz 2015). In summary, the first diagnostic criterion of AN requires that, "an individual voluntarily reduces and maintains an unhealthy degree of weight loss or fails to gain weight proportional to growth" (Sadock et al. 2015, p. 511).

Criterion B: Fear of Weight Gain

An intense fear of gaining weight or becoming fat, or behavior that prevents weight gain defines the second diagnostic criterion for anorexia nervosa (American Psychiatric Association 2013). Weight loss does not lessen weight-related concerns and instead may exacerbate fears about gaining weight (American Psychiatric Association 2013). The intense fear of becoming fat is often accompanied by a relentless drive for thinness, regardless of the adverse consequences of starvation (Sadock et al. 2015).

The intense fear of gaining weight is found in all patients with anorexia nervosa, a characteristic of the disorder that Sadock and colleagues (2015) suggest contributes to the lack of interest in or resistance to therapy that is commonly found among individuals with AN. The preoccupation with weight and shape becomes the dominating and consuming theme of thoughts, moods, and behaviors (Sadock et al. 2015). Individuals with anorexia nervosa fear weight gain, but are consumed by thoughts about food, which is evident in some of the behaviors that are characteristic of individuals with AN (e.g., collecting recipes, preparing meals for others, hiding food in peculiar places).

Criterion C: Bodily Disturbance & Self-Evaluations

The final diagnostic feature indicates the presence of a disturbed self-perception of body weight or shape that may influence evaluations of self or of the seriousness of the low body weight (American Psychiatric Association 2013). This may manifest as experiencing a global sense of being overweight, or as concerns regarding particular parts of the body as being "too fat" (American Psychiatric Association 2013, p. 340).

The importance attributed to thinness is considered to be a source of self-esteem and therefore, individuals with AN employ a wide-array of techniques to evaluate body size and shape (American Psychiatric Association 2013; Sadock et al. 2015). Such evaluations are based on weight- and body-checking behaviors, which include frequent weighing, pinching parts of the body, using the mirror to evaluate specific body parts (e.g., to check for perceived areas of fat), and obsessive measuring of body parts (e.g. measuring body dimensions with a tape measure) (American Psychiatric Association 2013; Sadock et al. 2015). The DSM 5 indicates that because the self-esteem of individuals with AN is highly dependent on the evaluations and perceptions that they have about their body shape and weight, weight loss is a notable achievement and indicative of self-control; weight gain is an undesirable failure that is interpreted as a lack of self-discipline (American Psychiatric Association 2013).

Subtypes

The DSM 5 differentiates between two subtypes of anorexia nervosa, the restricting type and the binge-eating/purging type (American Psychiatric Association 2013). The difference between the two subtypes is partially founded in the presence or absence of binge eating or purging behavior.

An episode of binge eating is a short period of time (e.g. within any 2-hour period) during which an individual consumes quantities of food that are larger than expected with consideration given to context and time, and is often associated with a perceived lack of control over the eating behavior; purging involves inappropriate compensatory behaviors intended to prevent weight gain and often manifest as self-induced vomiting or the misuse of laxatives, diuretics, or enemas (American Psychiatric Association 2013).

The restricting subtype of anorexia nervosa requires that no recurrent episodes of binge-eating or purging behaviors have occurred within the last three months (American Psychiatric Association 2013). Approximately half of all cases of anorexia nervosa fall under the food-restricting category (Sadock et al. 2015), and the DSM 5 notes that individuals diagnosed with this subtype lose weight through dieting, fasting, and excessive exercise (American Psychiatric Association 2013).

Food intake is rigidly constrained, typically limiting intake to less than 300 to 500 calories per day and no grams of fat (Sadock et al. 2015). Moreover, these individuals present as ceaselessly and compulsively overactive, with common activities including ritualistic exercising, extensive cycling, walking, jogging, and running (Sadock et al. 2015). Individuals diagnosed with restricting anorexia nervosa are often reported to have obsessive-compulsive characteristics that are related to food (Sadock et al. 2015).

The binge-eating/purging subtype specifies that within the past three months, recurrent episodes of binge-eating or purging behaviors have occurred (American Psychiatric Association 2013). This subtype typically presents as alternating attempts at rigorous dieting that are interrupted by intermittent periods of binge or purging episodes. According to Sadock and colleagues (2015), purging is a behavioral form of compensation for unwanted calories that is most commonly achieved through self-induced vomiting and laxative abuse, however diuretics and enemas may also be used, although less frequently. The DSM 5 suggests it is possible for these individuals to demonstrate patterns of purging behaviors without binge-eating (i.e., repetitive purging without prior binge eating or after ingesting a relatively small amount of food (Sadock et al. 2015; American Psychiatric Association 2013).

This subtype is frequently associated with substance abuse, impulse control disorders, and personality disorders, and individuals with this subtype also tend to have family members who are obese or overweight (Sadock et al. 2015). Additionally, when compared to individuals with the restricting subtype, individuals with the binge-eating/purging subtype have higher suicide rates and more frequently report personal histories of being overweight (Sadock et al. 2015).

Onset, Prevalence, & Affected Populations

Reports of anorexia nervosa have become more frequent in the past several decades, with the diagnosis most common among prepubertal girls (Sadock et al. 2015). The DSM 5 indicates that the 12-month prevalence among young females is approximately 0.4 percent (American Psychiatric Association 2013), however, Sadock and colleagues suggest that the disorder occurs in about 0.5 to 1 percent of adolescent girls. Onset most frequently occurs in the midteens, between ages 14 and 18, but research indicates that in up to five percent of anorectic patients onset occurs in their early 20s (Sadock et al. 2015, p. 509). Much less is known about the prevalence of anorexia nervosa among males, but the DSM 5 (American Psychiatric Association 2013, p. 341) estimates that anorexia nervosa is found in clinical populations with a 10:1 female-to-male ratio. This estimate is supported by Sadock et al. (2015), who indicate that the disorder occurs 10 to 20 times more frequently in females compared to males.

Cognitive-Behavioral Theory

Kreitler et al. (2003) indicate that many studies emphasize the biological, genetic, and physiologic causes of the anorexia nervosa, in addition to social, cultural and environmental etiological influences. Moreover, the researchers outline numerous theoretical frameworks and their respective explanations of individual psychological factors in anorexia nervosa, highlighting the need for a comprehensive theory of anorexia nervosa that integrates the principal motivations and the behavioral manifestations of the disorder.

Cognitive-behavioral theory may be conceptualized as a general term used to describe a set of interrelated theories that share fundamental assumptions, therapeutic techniques, and intervention strategies. Although both traditions maintain their respective models and theories, cognitive and behavioral theories have merged to produce an integrated theoretical perspective.

Cognitive-behavioral theory asserts that all human activity has three modalities: cognition, emotion, and behavior (McGuire 2000). These dimensions of human functioning are the products of the interchange between the internal world and the external environment. Thus, in the most rudimentary sense, the cognitive-behavioral framework addresses the relationship between thoughts, feelings, and actions.

However, an important aspect of cognitive-behavioral theory involves the fundamental distinction between "cognitive activity" and "behavior" (Hupp, Reitman, & Jewell 2008, p. 263). From both behavioral and cognitive perspectives, two types of behavior have been conceptualized. "Covert behavior" describes cognitive activities, the "behaviors that are performed in people's thoughts or cognitions" (Cobb 2008, p. 223). Feelings and emotions may be conceptualized as a component of covert behavior. In contrast, "overt behavior is composed of observable outward actions" (Cobb 2008, p. 223). Cognitive-behavioral theory emphasizes the interplay between covert behaviors, overt behaviors, and the environment, and asserts that both types of behavior are learned (Hupp et al. 2008).

Kendall (as cited in Hupp et al. 2008, p. 263) holds that cognitive-behavioral theory places the "greatest emphasis on the learning process and the influence of the models in the social environment, while underscoring the centrality of the individual's mediating/information processing style and emotional experiencing." Thus, a brief examination of both the cognitive and behavioral theoretical constructs is warranted in order to critically evaluate anorexia nervosa from a cognitive-behavioral perspective.

Cognitive Theory

Cognitive theories are structured around the principle that individuals' emotional reactions and behaviors are highly influenced by cognitions. Westbrook, Kennerley, and Kirk (2011) elaborate on this point and describe individuals' cognitions as the, "thoughts, beliefs and interpretations about themselves or the situations in which they find themselves – fundamentally the meaning they give to the events of their lives" (p. 4). When presented with similar events or situations, individuals respond in different ways, thus stressing the importance of the processes involved in interpreting and assigning meaning to environmental stimuli or events.

Covert behaviors give rise to various emotions, which suggests that people process information through a filter of their own thoughts and beliefs (Cobb 2008). For example, according to Beck and colleagues (1979), individuals with depression indicated high levels of critical and judgmental self-beliefs and indicated negative feelings, such as hopelessness and helplessness. Negative self-beliefs function as a filter through which individuals interpret environmental stimuli and events, thereby influencing perceptions of stimuli or interactions more negatively (Cobb 2008). Such overly negative perceptions develop into problematic beliefs (e.g., "change is impossible and life is hopeless"), subsequently leading to adverse emotions (Cobb 2008, p. 229-30). These problematic beliefs are commonly referred to in the literature as "cognitive errors" (Beck 1979) and "irrational beliefs" (Ellis 1977).

When the propositions and attributions that organize an individual's information processing contribute to the maintenance of distressing emotional states or dysfunctional behaviors, they become maladaptive. That is, the individual misinterprets or misperceives information that portrays a, "systematically disoriented picture of self and of the world" (McGuire 2000, p. 27).

Behavioral Theory

From a traditional behavioral perspective, human behavior is based on a model of reactivity, such that individuals are recipients of both external and internal stimuli and they respond to stimuli in either adaptive or maladaptive ways (Beck & Haigh 2014). Behavioral theories are primarily interested in learning, and individuals employ various processes that are important to maintain or alter behaviors (Robbins, Chatterjee, & Canda 2012). Both classical conditioning and operant conditioning are considered the main processes through which behaviors are learned. Classical conditioning suggests that environmental events or stimuli elicit specific behavioral responses. This process emphasizes learning that occurs on the basis of association of one stimulus with another, and is predominately concerned with the antecedents of behavior (Robbins et al. 2012).

In opposition to classical conditioning, operant conditioning focuses on reinforcement, rather than association, and is predominately concerned with the consequences of behavior. From this perspective, behavior is perpetually self-reinforcing, such that, "…the behavior is strengthened by its consequence and the consequence itself is the reinforcer for the behavior" (Robbins et al. 2012, p. 349). Two types of consequences exist, reinforcement and punishment, the former increasing the likelihood of the behavior and the later decreasing the likelihood of the behavior (Robbins et al. 2012). Reinforcement can be either positive or negative, however, both subtypes lead to a strengthened response. Positive reinforcement involves adding something to the environment. Whereas negative reinforcement increases a behavior and occurs when an aversive stimuli is removed from the environment, punishment weakens a behavior through either an application of an aversive consequence or the removal or a positive reinforcer (Robbins et al. 2012). These terms will become more relevant in their application to anorexia nervosa.

According to social learning theory, individuals may learn vicariously without the immediate experience of stimulus-response condition (e.g., the positive or negative consequences of behavior) (Cobb 2008). This theory presumes that individuals observe the behaviors of others, "capture cognitive images of behavior, and, in their thoughts, they replay or rehearse the behavior" (Cobb 2008, p. 222). When the behavior is enacted in the future, individuals anticipate reinforcement similar to that elicited in the original modeling of the behavior (Cobb 2008).

Central to social learning theory is the belief that behavior is based on the interaction between internal and external influences, with consideration given to cognition, thus underscoring the importance of cognitive processes in behavioral learning. Bandura (1977) assigns significant importance to the role of covert behaviors, such as thoughts and cognitions. He acknowledges that neither internal causes nor environmental stimuli determine human behaviors, but rather, "psychological functioning is accounted for by a reciprocal interaction of personal and environmental determinants" (p. 11-12). This "reciprocal determinism" offers individuals control in both cognitive and environmental contexts, which subsequently impacts behaviors (Robbins et al. 2012, p. 354). Thus, social learning theory offers a synthesis of cognitive, behavioral, and social and environmental elements, with a particular emphasis given to cognitive theory and principles of behavior modification (Robbins et al. 2012).

The Defining-Maintaining Paradox

Cognitive-behavioral theory provides a noteworthy lens to evaluate the DSM 5 diagnostic criteria for anorexia nervosa. The diagnostic criteria acknowledge that anorexia nervosa involves both behavioral and cognitive components, thus defining the disorder in terms of maladaptive behaviors and dysfunctional cognitive processes. However, the defining features of the disorder fail to elaborate on such cognitions and cognitive processes, which are arguably the fundamental mechanisms that cause the disorder to persist. In other words, the same conditions that define anorexia nervosa are paradoxically related to its maintenance. The following section offers a cognitive-behavioral account of anorexia nervosa that examines the underlying cognitive processes involved in the maintenance of the disorder.

General Overview

Kleifield, Wagner, and Halmi (1996) offer a cognitive-behavioral conceptualization of anorexia nervosa that is rooted in two core beliefs about the disorder, which are expanded on to provide a general overview of the disorder from a cognitive-behavioral perspective.

The first assumption indicates that the disorder develops as a coping strategy in response to unfavorable experiences, developmental transitions, and stressful life events, which often trigger adverse emotions. These negative emotions are relieved by "anorectic behaviors," such as dieting and other forms of weight-control, which indicates that individuals with anorexia nervosa may lack effective coping skills required to manage various components of their lives (Kleifield et al. 1996, p. 715). Successful food restriction and weight loss increase self-esteem through feelings of competence, confidence, and self-control, thus counteracting other negative emotions (Kleifield et al. 1996).

The second core belief is that strict dieting and weight control behaviors develop into fixed patterns of behavior (Kleifield et. al 1996). There exists a "self-fulfilling interplay between fear and avoidance," such that as food restriction and avoidance become more intense and rigid, hunger and nutritional deprivation grow, which results in a greater preoccupation with eating, shape, and weight (Kleifield et al. 1996, p. 715). The lives of individuals with anorexia nervosa become centralized around eating, shape, and weight with an increase in efforts to avoid overeating, "fatness," and weight gain (Fairburn, Cooper, & Shafran 2003, p. 510). "The more extreme their restrictions, the more extreme their fears and vice versa" (Kleifield, et al. 1996, p. 715). Thus, in the active pursuit of dietary control, thinness, and weight loss, individuals with anorexia nervosa must "redouble" efforts at food avoidance to alleviate their fears and anxiety (Kleifield et al. 1996, p. 715; Fairburn et al. 2003). That is, the central diagnostic feature of anorexia nervosa manifests as a fervent pursuit of thinness resulting from overwhelming fears of becoming fat (Gabbard 2014).

Cognitive-Behavioral Theory & Anorexia Nervosa

Fairburn, Shafran, and Cooper (1999) proposed a cognitive-behavioral theory of anorexia nervosa, and they later developed a "transdiagnostic" cognitive-behavioral theory of eating disorders (Fairburn et al. 2003). Moreover, Grave (2013) expands on the latter of the two theories offered by Fairburn and colleagues. These cognitive-behavioral approaches to anorexia nervosa provide a foundation to evaluate the diagnostic features of the disorder and elaborate on the maintaining mechanisms and processes of anorexia nervosa.

Cognitive elements are prominent in anorexia nervosa, and it is clear that individuals with eating disorders assign meaning to weight and body shape. More specifically, cognitive-behavioral models of eating disorders, according to Vitousek and Hollon (1990), share a common set of ideas: that these individuals "endow weight with rich connotations, equate their personal value with the shape of their bodies, and use the regulation of weight to subserve numerous functions in their lives" (p. 192).

The extreme concerns about body weight and shape are central in both the diagnosis of anorexia nervosa and the cognitive behavioral accounts regarding the maintenance of the disorder (Fairburn et al. 1999). Fairburn et al. (2003) note an important additional component in maintaining the disorder, that of perfectionism and the need for self-control. According to Shafran, Cooper, and Fairburn (2002), clinical perfectionism is, "the overdependence of self-evaluation on the determined pursuit of personally, self-imposed, standards in at least one highly salient domain, despite adverse consequences" (p. 778).

Research indicates that perfectionism often co-occurs with eating disorders (Shafran et al. 2002). Individuals with anorexia nervosa demonstrate a fear of failure (i.e., gaining weight) and a persistent pursuit of success (i.e., thinness and behaviors that interfere with weight gain), despite negative consequences (i.e., significantly low weight; American Psychiatric Association 2013, p. 338; Shafran et al. 2002). The negative consequences of the disorder, the "starvation symptoms" (Fairburn et al. 2003), are tolerated because self-evaluations are largely based on the pursuit and attainment of their goals (i.e., restriction of energy intake and the maintenance of a significantly low weight; American Psychiatric Association 2013, p. 338). The physiologic and psychological consequences of anorexia nervosa have important functions in maintaining the disorder that will be discussed below.

From a cognitive-behavioral perspective, the core psychopathology of anorexia nervosa is an overvaluation of body shape and weight, as well as control in and over these three domains (Grave 2013; Fairburn et al. 1999, 2003). Many of the diagnostic features of anorexia nervosa are rooted in this core psychopathology, thereby suggesting that the shape/weight/eating/control overvaluation is critical in the maintenance of the disorder.

The shape/weight/eating/control overvaluation manifests as a maladaptive system for making self-judgments and evaluating self-worth (Grave 2013). Individuals with anorexia nervosa usually adopt a strict diet that is characterized by highly rigid and extreme dietary rules (Grave 2013), and these rules are often dichotomous in nature; they are either achieved or not achieved (Shafran et al. 2002). Such rigid dietary rules typically dictate the amount eaten, as well as the types of food consumed and where eating takes place (Fairburn et al. 1999).

Maintenance Mechanism 1

The restriction of energy intake is not only a criterion for diagnosis, but also presents as one of the key maintenance mechanisms of the disorder (American Psychiatric Association 2013; Fairburn et al. 1999). That is, dieting behaviors reinforce and intensify the core psychopathology of anorexia nervosa (Grave 2013). The restriction of food intake is a, "positive, cognitive reinforcement associated with the feeling of controlling shape, weight, and eating" (Grave 2013, p. 24). Thus, the ability to maintain control over shape/weight/eating becomes an index for self-control and self-worth, such that it is the act, rather than the outcome, that is an indicator of success (Fairburn et al. 1999). However, research suggests that the consequences of pursuing dietary restriction may further encourage preoccupations with shape/weight/eating/control in individuals with anorexia nervosa (Shafran et al. 2002; also see Keyes, et al. 1950).

Grave (2013) notes, "successful adherence to extreme and rigid dietary rules produces a condition of dietary restriction and being underweight, which is in turn associated with severe physical and psychosocial consequences" (p. 25). In the pursuit of shape/weight/eating/control, individuals with anorexia nervosa not only achieve a low body weight, but the low body weight produces a "starvation syndrome" that reinforces the overvaluation of eating/shape/weight/control, and thus present as a second mechanism through which anorexia nervosa is maintained (Fairburn et al. 2003, p. 521).

Maintenance Mechanism #2

The consequences of severe weight loss and restricted energy intake have important implications in the maintenance of anorexia nervosa. Fairburn et al.

(1999) suggest some of the symptoms may further promote dietary restriction by undermining self-control. For example, intense hunger may be interpreted as a threat to self-control over shape/weight/eating. Fairburn et al. (2003) further elaborate on the aspects of starvation that are crucial in the maintenance of the disorder. They note the symptom of social withdrawal, which isolates individuals with anorexia nervosa from potential resources and external influences that may be important in helping to reduce their eating/shape/weight/control overvaluation. Fairburn and colleagues also indicate that social isolation may promote "self-absorption" (p. 521).

Moreover, some of the starvation symptoms may be perceived in more dichotomous terms. Whereas some symptoms indicate success that an individual is maintaining control over his or her eating/shape/weight, other symptoms may be construed as evidence of failure.

Shafran et al. (2002) note that individuals with anorexia nervosa make self-evaluations through biased information processing, which includes biases in cognitive processes and information processing. The literature indicates that individuals with anorexia nervosa have organized cognitive structures that "unite views of the self with beliefs about weight" (Vitousek & Hollon 1990, p. 191). Such cognitive structures that are centralized around information pertaining to body size, shape, and weight are referred to in the literature as the "body self-schema" (Williamson, White, York-Crowe, & Stewart 2004). The body self-schema is presumed to influence the reactions, perceptions, thoughts, and behaviors in anorectic individuals; it directs attention to stimuli that are associated with their body and food, and biases the "interpretations of self-relevant events in favor of fatness interpretations" (Williamson et al. 2004, p. 714).

According to cognitive theory, the cognitive biases that occur as a result of the body self-schema in individuals with anorexia nervosa are likely to be activated by various stimuli that Williamson and colleagues (2004) elaborate on. These stimuli include: body- or food-related information, ambiguous stimuli, and situations that require the person to reflect on his or her body and eating patterns. The information errors in the cognitive processes of individuals with anorexia nervosa contribute to the maintenance of their dysfunctional eating behaviors.

Maintenance Mechanism #3

Attention bias involves cognitive processes through which individuals with eating disorders differentially attend to stimuli that are related to body weight and shape, as well as food and eating. Williamson et al. (2004) indicate that individuals with anorexia nervosa are more likely to attend to information that is consistent with the negative self-beliefs (stimuli associated with fatness) and to ignore information that may contradict these beliefs (stimuli associated with thinness). Thus, they are more likely to attend to actual or perceived errors, which manifests as hypervigilant monitoring for failures in controlling their diet or maintaining their body shape and weight.

Hypervigilent monitoring often involves both overt and covert checking behaviors (Shafran et al. 2002). In individuals with eating disorders, body-checking behaviors are engaged in as a means to gauge if their body shape has changed or if they have gained weight. However, body checking frequently results in the detection and magnification of perceived flaws and imperfections, thereby increasing bodily dissatisfaction and reinforcing the need for more rigid and extreme dietary restrictions (Grave 2013). This frequent and selective attention to performance often includes repeated calorie counting, monitoring of performance, and frequent shape and weight checking (Shafran et al. 2002).

Individuals with anorexia nervosa closely monitor their weight, and frequent weighing often results in concerns about insignificant fluctuations in weight, thereby maintaining the distorted beliefs about weight (Fairburn et al. 1999). Minor decreases in weight serve as powerful reinforcement, whereas weight gain (or no significant weight loss) may be construed as confirmation of poor self-control. In the case of the latter, the disorder is maintained through an intensification of efforts to restrict energy intake. Moreover, shape checking often magnifies perceived bodily imperfections. Through confirmatory information processing biases, such hypervigalent checking produces a cycle of perceived failures to control body shape and weight that influence the persistence of dietary restrictions. Fairburn and colleagues describe this process: "This hypervigilant checking in turn increases arousal, self-focused attention and anxiety, which encourages further monitoring, thereby establishing a vicious cycle" (p. 7).

In some cases, individuals may gradually feel that they cannot tolerate such self-scrutiny, thus prompting a decrease in the body checking behaviors (Fairburn et al. 1999). "Shape avoidance of the acts to maintain body dissatisfaction and therefore the eating disorder itself, because concerns and fears about shape and appearance tend to persist in the absence of a real idea of what one actually looks like" (Grave 2013, p. 33). Typically, these behaviors include avoiding mirrors, eschewing bathroom scales, refusing to expose the body to others, and wearing baggy clothes (Grave 2013). Consequently, without the means to disconfirm beliefs about their body, such avoidance behavior maintains the overvaluation of body/weight/shape/control

When failure is perceived, individuals may become increasingly reliant upon additional weight control techniques, including excessive exercise, self-induced vomiting, and the misuse of laxatives (Fairburn et al. 1999). Rigid exercise regimes may directly increase an individual's sense of self-control, thus reinforcing body/weight/shape/control overvaluation. However, extreme exercising, as well as self-induced vomiting and the misuse of laxatives, are behaviors that may stimulate a relaxation over control of eating. That is, individuals with anorexia nervosa believe in the effectiveness of these behaviors in body/shape/weight/control, thus minimizing their sense of control. Experiences of being without self-control prompt the individual to lose control over food intake, thus increasing the propensity to binge (Grave 2013).

A Note Regarding Onset, Adolescence, & The Media

Although this paper seeks to address the cognitive processes involved in the maintenance of the disorder, it is important to acknowledge that typical cases of anorexia nervosa have onset in adolescence (Grave 2013). Moreover, the central diagnostic feature of anorexia nervosa is the fervent pursuit of thinness resulting from an overwhelming fear of becoming fat. It is fair to question the role that the media has in advancing this ideal of thinness.

In Westernized societies, the media often portrays thinness as a measure of attractiveness, thus creating an unrealistic, idealized image of beauty. This standard of beauty is perpetually enforced and reinforced through societal standards and practices, thereby influencing young women's perceptions and evaluations of themselves and others.

Adolescence is a time during which individuals begin to develop an understanding of self in relation to others, and exposure to this popularized culture of thinness has a lasting effect on adolescent girls' self-confidence, self-esteem, and overall feelings of competency. Although an examination of the relationship between mass media, body image, and self-esteem is beyond the scope of this paper, it is an important point to acknowledge given that it is during adolescence when anorexia nervosa typically develops, and that it is a disdisorder that is most common among adolescent females.

Conclusion

The diagnostic criteria for anorexia nervosa in the *Diagnostic and Statistical Manual of Mental Disorders* (American Psychiatric Association 2013) acknowledge that anorexia nervosa involves both behavioral and cognitive components. Thus, the DSM 5 defines the disorder in terms of maladaptive behaviors and dysfunctional cognitive processes. Vitousek (1996) summarizes the cognitive-behavioral approach to anorexia nervosa:

> Reduced to its essence, the cognitive-behavioral model holds that anorexic and bulimic symptoms are maintained by a characteristic set of overvalued ideas about the personal implications of body shape and weight. These attitudes have their origins in the interaction of stable individual characteristics (such as perfectionism, asceticism, and difficulties in affect regulation) with sociocultural ideals for female appearance. Once formed, the beliefs influence the individuals who hold them to engage in stereotypic eating and elimination behaviors, to be responsive to eccentric reinforcement contingencies, to process information in accordance with predictable cognitive biases, and, eventually, to be affected by physiological sequelae that also serve to sustain disordered beliefs and behaviors (p. 384).

This paper employs cognitive-behavioral theory to examine the diagnostic criteria, contending that the defining features of the disorder are also the fundamental mechanisms that underlie the persistence of the disorder.

With consideration given to the maintaining processes as they relate to the disorder, three central maintaining mechanisms of anorexia nervosa were explored. The mechanisms, and thus the diagnostic criteria, are summarized as follows: 1) The *restriction of caloric intake* prompts behaviors that reinforce the *fear of fatness* and intensify the overvaluation of body/shape/weight/control; 2) Aspects of starvation and maintaining a *significantly low body weight* further encourage rigid dietary restrictions; 3) The *self-evaluations* dependent on weight and body checking behaviors feed the overvaluation of body/shape/weight/control and further encourage rigid dietary restriction.

Cognitive-behavioral theory offers a critical lens to examine the underlying cognitive processes involved in the maintenance of the disorder. Thus, from a cognitive-behavioral framework, it appears that it is the same conditions that define anorexia nervosa that are paradoxically related to its maintenance.

References

American Psychiatric Association. 2013. *Diagnostic and Statistical Manual of Mental Disorders (DSM 5)* (5th ed.). Washington, DC: American Psychological Association.

Bandura, A. 1977. *A Social Learning Theory*. Englewood Cliffs, NJ: Prentice-Hall.

Beck, A., Rush, A., Shaw, F., & Emery, G. 1979. *The Cognitive Therapy of Depression*. New York, NY: Guilford Press.

Beck, A. & Weishaar, M. 2008. Cognitive therapy. In R. Corsini & D. Wedding (eds.), *Current Psychotherapies* (9th ed.), 276-309. Belmont, CA: Brooks/Cole.

Cobb, N. 2008. Cognitive-behavioral theory and treatment. In N. Coady & P. Lehmann (eds.), *Theoretical Perspectives for Direct Social Work Practice: A General-Eclectic Approach* (2nd ed.), 221-248. New York, NY: Spring Publishing Company.

Ellis, A. 1977. Rational-emotive therapy: Research data that supports the clinical and personality hypotheses of RET and other mods of cognitive-behavior therapy. *The Counseling Psychologist, 7*(1), 2-42.

Fairburn, C., Shafran, R., & Cooper, Z. 1999. A cognitive behavioural theory of anorexia nervosa. *Behaviour Research and Therapy, 37*(1), 1-13.

Fairburn, C., Cooper, Z., & Shafran, R. 2003. Cognitive behaviour therapy for eating disorders: A "transdiagnostic" theory and treatment. *Behaviour Research and Therapy, 41*(5), 509-528.

Gabbard, G. 2014. *Psychodynamic Psychiatry in Clinical Practice*. Washington, DC: American Psychiatric Publishing.

Garner, D. & Bemis, K. 1982. A cognitive-behavioural approach to anorexia nervosa. *Cognitive Therapy and Research, 6*(2), 123-150.

Grave, R. 2013. *Multistep Cognitive Behavioral Therapy for Eating Disorders: Theory, Practice, and Clinical Cases*. Lanham, MD: The Rowman & Littlefield Publishing Group, Inc.

Hupp, S., Reitman, D., & Jewell, J. 2008. Cognitive-behavioral theory. In M. Hersen & A. Gross (eds.), *Handbook of Clinical Psychology Volume 2: Children and Adolescents*, 263-287. Hoboken, NJ: John Wiley & Sons.

Keyes, A., Brozek, J., Henschel, A., Mickelsen, O., & Taylor, H. 1950. *The Biology of Human Starvation Volume 2*. Minneapolis, MN: University of Minnesota Press.

Kleifield, E., Wagner, S., & Halmi, K. 1996. Cognitive-behavioral treatment of anorexia nervosa. *Psychiatric Clinics of North America, 19*(4), 715-737.

Kreitler, S., Bachar, E., Canetti, L., Berry, E., & Bonne, O. 2003. The cognitive-orientation theory of anorexia nervosa [Abstract]. *Journal of Clinical Psychology, 59*(6), 651-671.

McGuire, J. 2000. *Cognitive-Behavioural Approaches: An Introduction to Theory and Research*. London: Home Office Communication Directorate.

Robbins, S., Chatterjee, P., & Canda, E. 2012. *Contemporary Human Behavior Theory: A Critical Perspective for Social Work* (3rd ed.). Saddle River, NJ: Allyn & Bacon.

Sadock, B., Sadock, V., & Ruiz, P. 2015. *Kaplan and Sadock's Synopsis of Psychiatry: Behavioral Sciences/Clinical Psychiatry* (11th ed.). Philadelphia, PA: Wolters Kluwer.

Shafran, R., Cooper, Z., & Fairburn, C. 2002. Clinical perfectionism: A cognitive-behavioural analysis. *Behaviour Research and Therapy, 40*(7), 773-791.

Vitousek, K. 1996. The current status of cognitive-behavioral models of anorexia nervosa and bulimia nervosa. In P. Salkovskis (ed.), *Frontiers of Cognitive Therapy*, 383-418. New York, NY: Guildford Press.

Vitousek, K. & Hollon, S. 1990. The investigation of schematic content and processing in eating disorders. *Cognitive Therapy and Research, 14*(2), 191-214.

Westbrook, D., Kennerley, H., & Kirk, J. 2011. *An Introduction to Cognitive Behavior Therapy: Skills and Applications* (2nd ed.). Thousand Oaks, CA: Sage Publications.

Williamson, D., White, M., York-Crowe, E., & Stewart, T. 2004. Cognitive-behavioral theories of eating disorders. *Behavior Modification, 28*(6), 711-738.

A Feminist Perspective on Major Depressive Disorder in Men

Emily Morris

If we look at the ways the systems on the mezzo, macro, and micro levels of social work interact, we can see how environmental factors, policies, and institutions perpetuate inequalities between different groups and cause problems for individuals. Fortunately, advocates often band together to work towards equal opportunities for members of all races, classes, and backgrounds. Most recently, gender equality has been in the spotlight as an important goal, not only in the U.S. but globally. In fact, UN Women, an organization created by the United Nations General Assembly, just announced a gender equality campaign called, "He for She" that aims to improve the lives of women by including both men and women in the work ("He for She," 2015).

When Emma Watson, a British actress and the Goodwill Ambassador for UN Women, announced the campaign back in September 2014, she made the important point that we cannot achieve gender equality if only women work towards it; the issue inherently calls for men to be included in the fight too (Cole, 2015). Watson argued that up until now, men have not been invited to join the feminist movement, which is problematic, given that gender inequality negatively affects both genders (Cole, 2015). For example, both men and women are imprisoned by gender role stereotypes, as Watson explained:

> Because to date, I've seen my father's role as a parent being valued less by society, despite my need of his presence as a child as much as my mother's. I've seen young men suffering from mental illness, unable to ask for help for fear it would make them less of a man. In fact, in the UK, suicide is the biggest killer of men between [ages] 20 to 49, eclipsing road accidents, cancer, and coronary heart disease. I've seen men made fragile and insecure by a distorted sense of what constitutes male success. Men don't have the benefits of equality, either (Cole, p. 1, 2015).

As Watson implies, gender inequality affects individuals of both genders, yet we most often look at how it affects women. This makes sense, given that women have historically had less power than men across many cultures; however, without also understanding the ways in which we are failing our men, we are blinding ourselves to a key contributor to the gender inequality problem.

Currently, in the U.S. men complete suicide four times more often than women ("Facts and Figures," 2015). Even though women are twice as likely to be diagnosed with major depressive disorder and also attempt suicide more often than men, men are four times more likely to complete suicide (American Psychiatric Association, 2013). If so many fewer men are diagnosed with major depressive disorder, then why are so many completing suicide? Using a feminist ecological perspective to tackle this question, I am going to argue that the social and political systems in the U.S. work together to provide more mental health services for depression to women than men, and to exclude men from treatment opportunities. First, I will outline the tenets of feminist ecological theory and how one can view mental health issues through a feminist lens. Then, I will briefly explain the clinical manifestation of major depressive disorder. I will go on to break down these two key systems with which individuals in the U.S. interact on a daily basis in order to explain how they encourage women to seek mental health services for depression and discourage men from doing so.

Feminist Ecological Theory

Feminism is the idea that men and women should have equal opportunities and rights (*Merriam-Webster Dictionary*, 2015). Within this broad framework, feminist theory asserts that the identities we construct are influenced by our relationships and environmental factors and, "that gender oppression is a social fact that operates to produce inequalities against women in the social, cultural, political, and economic domains" (Ballou, Matsumoto, & Wagner, p. 102, 2002). When looking at situations through a feminist lens, one explores the power distribution in relationships, particularly in relation to gender. Feminist ecological theory uses feminism as a basis to argue that all of the systems we interact with daily, including the interpersonal, social, spiritual, political, economic, environmental, cultural, and educational, affect our actions, perception, emotional and cognitive development, as well as the construction of our identities (Ballou, Matsumoto, & Wagner, 2002). The authors of this model aim to demonstrate the complex and multiple forces that affect human nature and development by combining ideas from feminist therapy theory, critical theory, liberation psychology, ecopsychology, and transformative multiculturalism (Ballou, Matsumoto & Wagner, 2002). To demonstrate that men receive fewer opportunities to receive mental health services for depression than women, I will use feminist ecological theory to analyze how the social and political systems interact to produce this iniquity.

The Social or Interpersonal Systems

For the purposes of this analysis, social and interpersonal systems include friends and family, teachers, coworkers, community groups, the school system, and any relationships between people. In the feminist ecological model, this is called the microsystem (Ballou, Matsumoto, & Wagner, 2002). Using a feminist lens, we can see how individuals' self-esteem, perspective, and emotional development are affected by the beliefs and expectations that others have for them (Ballou, Matsumoto, & Wagner, 2002). Feminist ecological theory asserts that our gender influences our interactions with others, the development of our belief systems, and the expectations that we have for our daily lives, including our career, education, and community involvement (Ballou, Matsumoto, & Wagner, 2002).

Basically, what we learn during the socialization process, completed by our parents and peers' influences on us, differs based on our gender. Women are taught, "to be nonassertive, self-sacrificing, and dependent on others…" while men are taught to be independent, physically tough, competitive, assertive, and emotionally stoic (Sparks, p. 286, 2002; Addis, 2008). Women are taught to explore their feelings, while men are not; instead, they often learn by example to suppress or externalize their emotions to appear more masculine (Addis, 2008).

Much of the research that shows that gender-role orientation does affect one's susceptibility to depression looks solely at female gender roles that contribute to this diagnosis (Addis, 2008). If we think about what is defined as traditionally masculine, including emotional stoicism, aggression, and fearlessness, we can see the incentive for men to avoid seeking mental health services or advocating for themselves in that arena (Addis, 2008). Even though many U.S. citizens believe in gender equality, we still learn gendered roles from our family experiences, and we often perpetuate these stereotypical roles without realizing that we are doing it. One example is the traditional, gendered division of labor in many families, where the female partner takes on the primary caregiver role and the male partner acts as breadwinner (Ballou, Matsumoto, & Wagner, 2002). If one sees this division as a child, one may fall into or choose a similar role as an adult.

Given the pressure on the male partner to financially support the family, and the underlying association of a stoic persona as successfully masculine, one can see why it would be demeaning to be labeled as mentally ill or depressed. The clinical picture of a person with major depressive disorder does not fit with the image of a traditionally masculine man. While women build their self-esteem around their relationships with others, men build their self-esteem based on their performance and productivity (Sparks, 2002). Major depressive disorder is characterized by depressed mood, fatigue, difficulty concentrating, feelings of worthlessness, and a lack of interest in one's former hobbies; further, these symptoms impair one's daily functioning (American Psychiatric Association, p. 161, 2013). All of these symptoms represent a lack of central masculine values like confidence, productivity, and the ability to accomplish tasks at work and at home. Clearly, it is not in a man's best interest to identify as depressed.

If more men are depressed than we realize, then, what do their symptoms look like? In his article, "Gender and Depression in Men," Michael Addis reviews various research on male depression, writing that, "Because masculine gender norms generally encourage action and discourage introspection, men who are depressed and affected more strongly by such norms are hypothesized to exhibit more externalizing symptoms" (Addis, p. 157, 2008). These externalizing symptoms result in disorders like substance abuse, antisocial personality disorder, and anger management issues (Addis, 2008). These externalizing symptoms have been referred to as a "masculine [form of] depression" (Addis, p. 157, 2008). Men may be coping poorly with depression, causing these externalizing issues, or even somatic symptoms (Addis, 2008). At the same time, since women are often socialized to explore their emotions while men are not, men may have difficulty identifying their depressed feelings and may choose to distract themselves with a task instead (Addis, 2008). Finally, some men may suppress their emotions in order to adhere to the emotional stoicism associated with successful masculinity instead of externalizing their feelings (Addis, 2008).

Given that men are also socialized to be independent and therefore to avoid asking for help, fewer men also seek mental health services than women (Addis, 2008). In the U.S., men are taught to cope in ways that prevent them from expressing depression in the "typical" ways outlined in the DSM; as a result, they may suppress or externalize their emotions via problematic behaviors. As one author explains, men and women react differently to stressors due to their unique socialization: "...the differences in the ways that boys and girls react to stressful situations may be due to such factors as emotional expressivity, socialization about appropriateness of coping strategies, or experience with a range of coping opportunities" (Sparks, 2002). The ways in which men cope with what may be labeled as depression may not look like traditional depression according to the *Diagnostic and Statistical Manual of Mental Disorders*, causing fewer men to be diagnosed than women.

The Political System

Within the current political system in the U.S., many groups are advocating for women's rights, especially those related to healthcare. Given the complexity of reproductive issues and history of unequal rights between the genders, this work is especially important. However, as we have created more and more policies to support women's physical and mental health, we have left men by the wayside. Due to the implementation of policies that serve women in combination with a socialization process that equates emotional stoicism with masculinity, we have failed to encourage men to seek mental health services and educate them on when and how to do so.

For over a decade, women's health has been in the spotlight in the U.S. In fact, the Patient Protection and Affordable Care Act, passed in 2010 by President Obama, includes parts directed specifically at improving health services for women (Sianko, 2011). The Mothers Act, for example, provides specific funding for services related to post-partum mood disorders and other illnesses, as well as encourages research and national education in this area (Sianko, 2011). Although this is a necessary and extremely beneficial policy for women, there are no equivalent policies to address men's mental health, even though the suicide statistics point to a need for intervention. Another example of the political focus on women's mental health is President Barack Obama's appointment of a White House Advisor on Violence Against Women, a position that did not exist before his presidency (Sianko, 2011). President Obama also created "the White House Council on Women and Girls to advise the Administration on child care, parental leave, family violence, and many other issues of particular interest to women" (Sianko, p. 170, 2011). Clearly these political moves have the potential to benefit many women, children, and families; however, these actions send the message to men that their mental health is not as important as women's, a message that many are already receiving through the socialization process.

These political moves in the U.S. reflect a global focus on women's health that began "more than a decade ago, when the World Health Organization (WHO) divisions on mental health and family health released a groundbreaking report...[that] launched a global movement to improve women's mental health" (Sianko, p. 168, 2011). What followed were a number of key events, conferences, and research done on women's mental health, including the Summit on Women and Depression put on by the American Psychological Association in 2000 and the First World Congress on Women's Mental Health put on by various organizations in Berlin in 2001 (Sianko, 2011). Even before this, though, U.S. researchers were increasingly focusing on women's mental health: in 1993, for example, "the reauthorization of the National Institutes of Health included the authority to create the Office of Research on Women's Health" (Sianko, p. 168, 2011). This office, which was originally established in 1990, promotes and regulates research done on women's physical and mental health in order to improve services to women ("History and Mission," 2015).

Looking at President Obama's recent political moves in relation to women's health, we can see how the global gender equality movement, specifically the movement aiming to improve women's mental health, affected policy in the U.S. and effectively led to increased education, awareness, and action surrounding women's mental health. At the same time, these huge gains for women contrast starkly with the lack of mental health education, awareness, and action on men's mental health. One study published in the *Journal of Women's Health and Gender-Based Medicine* found that women with high scores on the Beck Depression Inventory (BDI) were more likely than men with high BDI scores to be diagnosed with depression by their primary care doctor (Bertakis et. al., 2004). This means that the difference in diagnosis rate was not caused by a "male depression" that looks different than traditional depression; the same scale was used for both men and women, and those who scored higher were more likely to be diagnosed with depression if they were female (Bertakis et. al., 2004).

Looking at this finding through the lens of feminist ecological theory, we can see how the interaction between the healthcare system, the political system, social systems, and the media could cause such a variation in diagnosis rates. Policies like those listed above encourage research and funding for women's mental health studies and services, but discourage the equivalent for men. These policies are discussed widely in the media, including their basis, which is the gender gap in diagnosis of depression. Unfortunately, there is evidence that this gap is narrowing (Addis, 2008), and we have yet to create a council or advisor on men's mental health, or cover it as widely in the media. Moreover, the media affects citizens' perceptions of current issues and can therefore influence diagnosis rates; in other words, even when seeing a male and female patient with similar BDI scores, doctors may be more likely to diagnose the woman with major depressive disorder due to internal biases caused by one's socialization and the messages one has received from the media.

The authors of this study also found that the more frequently participants visited their doctor, the more likely they were to be diagnosed with depression, regardless of gender; further, women visited the doctor more often than men (Bertakis et. al., 2004). A similar study that looked at the effects of gender on help-seeking attitudes found that negative views of psychological openness contributed to the fact that men were less likely to seek mental health services (Mackenzie, 2006). In other words, through the socialization process, boys are often taught that expressing one's feelings is a form of weakness, and that successful men are not weak. As a result, they may view psychological openness as a negative quality and consequently seek mental health services less often. Another study that explored the perception of mental health services based on gender found that women hold more positive attitudes towards mental health services than men and suggested that mental health providers work to alleviate men's fears about being labeled as, "weak" or "crazy" or "dependent" due to their mental health status (Currin, Hayslip, & Temple, p. 337, 2011). Through this analysis, we can see how the political agenda affects the healthcare system and the access to, education on, and quality of mental health services for male individuals; moreover, the socialization process, including one's education and family experiences, among other relationships, shape one's coping skills, perception, and use of mental health services.

Why Focus On Men With Major Depressive Disorder?

There are a number of reasons I chose to focus on men with major depressive disorder, many of which are woven into the above paragraphs, the main one being that men are four times more likely than women to complete suicide ("Facts and Figures," 2015). Although factors other than mood can contribute to suicide, major depressive disorder increases one's risk for suicide, and suicidality is written into the criteria for the disorder: "Recurrent thoughts of death (not just fear of dying), recurrent suicidal ideation without a specific plan, or a suicide attempt or a specific plan for committing suicide" is noted as one of nine potential symptoms (American Psychiatric Association, p. 161, 2013). Others include feelings of worthlessness, hopelessness, fatigue, and decreased pleasure in one's hobbies (American Psychiatric Association, 2013), all of which could contribute to suicidal ideation or attempts.

The fact that men are completing suicide so much more frequently than women, but are diagnosed with depression so much less, points to a problem in the access to services, education on mental health, the diagnosis process, or maybe all of the above. One study that looked at gender differences in the rates of exposure to stressful life events and sensitivity to the effects of these events found that neither gender experiences stressful events more frequently than the other and neither is uniquely sensitive to the stressors that they do experience (Kendler, 2001). This means that something other than the frequency or nature of life stressors is causing the gap in diagnoses between men and women; could it be that more men than we know are experiencing depression but not receiving treatment?

In the U.S., the political focus on women's mental health over the last twenty years has led to increased research on and funding for women's mental health services, which has in turn affected our healthcare system, including services offered, doctors' perceptions of mental health, and diagnosis rates; further, each individual's upbringing shapes one's perception of depression and his or her attitude towards mental health services. With a dominant socialization process that teaches young men to externalize or suppress their emotions, seeking therapy for one's emotional state does not fit into the idealized picture of traditionally successful masculinity. These political and social factors work together to create an environment that encourages women to seek mental health services for depression and discourages men from doing so. One author argues that almost no one has studied the gendered nature of our mental health system from a male perspective, making men's mental health issues effectively invisible in the social scientific community. This lack of research leads to a lack of funding, advocacy, and awareness, most likely causing negative consequences for men, women, children, and communities (Addis, p. 154, 2008).

Conclusion

As previous authors have argued (ie. Mackenzie, 2006), the gender gap in diagnosis of depression points to a need for more research and education on men's mental health issues. Although the recent changes in the healthcare system in the U.S. have increased the funding for and research on mental health services provided to women, it has left men without adequate support in this area. If we view this issue through feminist ecological theory, we can see how systems on the micro, macro, and mezzo levels work together to produce this iniquity. The traditional socialization process in the U.S., completed through one's early family and relationship experiences, teaches men that emotional stoicism, suppression, or externalization constitute male success, along with independence, which encourages men not to seek services. On the macro level, a national political agenda favoring research and funding for women's health also discourages men from seeking mental health services, sending the message that their mental health is less important than women's.

On the mezzo level, the global movement for gender equality, specifically in regards to improving mental health services for women, affects our national policy in the U.S., which then affects the healthcare system, what is published in the media, and the perception of consumers. Going forward, more research on men's mental health and their perception of depression specifically needs to be done. At the same time, we need more education on the status of men's mental health and widespread awareness regarding the gap in suicide completion. Given that men may exhibit or experience depression differently than women, only once detailed research on this topic is done can we design effective mental health services for men. In our fight for gender equality, as Emma Watson noted in her recent speech at the U.N. General Assembly, it is imperative that we include both genders in our work (Cole, 2015).

References

Addis, M. (2008). Gender and Depression in Men. Clinical Psychology: Science and Practice, 15(3), 153-168. Retrieved April 20, 2015, from Loyola Libraries.

American Psychiatric Association. (2013). *Diagnostic and statistical manual of mental disorders* (5th ed.). Washington, DC: American Psychiatric Association.

Ballou, M., Matsumoto, A., & Wagner, M. (2002). Toward a Feminist Ecological Theory of Human Nature. In *Rethinking Mental Health and Disorder: Feminist Perspectives* (pp. 99-143). New York, NY: The Guilford Press.

Cole, N. (n.d.). Full Transcript of Emma Watson's Speech on Gender Equality at the UN. Retrieved April 20, 2015, from http://sociology.about.com/od/Current-Events-in-Sociological-Context/fl/Full-Transcript-of-Emma-Watsons-Speech-on-Gender-Equality-at-the-UN.htm

Currin, J. B., Hayslip, B. J., & Temple, J. R. (January 01, 2011). The relationship between age, gender, historical change, and adults' perceptions of mental health and mental health services. *International Journal of Aging & Human Development, 72,* 4, 317-41.

Facts and Figures. (2015, January 1). Retrieved April 20, 2015, from https://www.afsp.org/understanding-suicide/facts-and-figures

Feminism. (2015, January 1). Retrieved April 20, 2015, from http://www.merriam-webster.com/dictionary/feminism

Gabbard, G. (1994). *Psychodynamic Psychiatry in Clinical Practice* (5th ed.). Washington, DC: American Psychiatric Press.

History and Mission. (2015, March 26). Retrieved April 20, 2015, from http://orwh.od.nih.gov/about/mission.asp

Home - HeForShe. (2014, January 1). Retrieved April 19, 2015, from http://www.heforshe.org

Kendler, K. S., Thornton, L. M., & Prescott, C. A. (January 01, 2001). Gender differences in the rates of exposure to stressful life events and sensitivity to their depressogenic effects. *The American Journal of Psychiatry, 158,* 4, 587-93.

Bertakis, K.D., Helms, J.L., Callahan, E. J., Azari, R., Leigh, P., & Robbins, J.A. (September, 2001). Patient gender differences in the diagnosis of depression in primary care. *Journal of Women's Health & Gender-Based Medicine.* 10(7): 689-698. doi:10.1089/15246090152563579.

Mackenzie, C. S., Gekowski, W. L., & Knox, V. J. (2006). Age, gender, and the underutilization of mental health services: The influence of help-seeking attitudes. Aging and Mental Health, 10, 574-582.

Sianko, N. (2011). Gender Equality and Women's Mental Health: What's on the Agenda? *American Journal of Orthopsychiatry, 81*(2), 167-171. Retrieved April 20, 2015, from Loyola Libraries.

Sparks, E. (2002). Depression and Schizophrenia in Women: The Intersection of Gender, Race/Ethnicity, and Class. In *Rethinking Mental Health and Disorder: Feminist Perspectives* (pp. 279-305). New York, NY: The Guilford Press.

Swartz, S. (2013). Feminism and psychiatric diagnosis: Reflections of a feminist practitioner. *Feminism & Psychology, 23*(1), 41-48. Retrieved April 20, 2015, from Loyola Libraries.

Narcissistic Personality Disorder through a Social Psychology Perspective

Divya Oommen

Introduction

In today's society, admiration and praise are often sought out or aspired for. It is displayed through social media, television, peers and family. Essential features of Narcissistic Personality Disorder (NPD) are the need for admiration and the lack of empathy (APA, 2013). Social Psychology looks at how the norms, values, and interactions with others influence our behavior. The environment to which we are exposed plays a substantial role in shaping our personalities. Factors in the environment which shape our personality are: the culture in which we are raised, our early conditioning, the norms among our family, friends and social groups, and other influences that we experience (Smith, Mackie & Claypool, 2014). Social Psychology is the scientific study of the effects of social and cognitive processes on the way individuals perceive, influence or relate to others (Smith, Mackie & Claypool, 2014). Through the lens of social psychology, we will review the diagnosis of Narcissistic Personality Disorder to make a judgement of its placement in the DSM 5.

Narcissistic Personality Disorder: The Diagnosis

In the Fifth Edition of the *Diagnostic and Statistical Manual of Mental Disorders* (DSM 5), the section on Personality Disorders has 10 specific disorders. The definition of personality disorders requires a pattern of inner experience and behavior that deviates from the expectations of the individual's culture. It is pervasive, inflexible, and has an onset in adolescence or early adulthood impairment (American Psychiatric Association [APA], 2013). It is stable over time and leads to distress or impairment. One of these disorders is Narcissistic Personality Disorder, which is a pattern of grandiosity, need for admiration, and lack of empathy (APA, 2013).

There are three clusters which categorize the personality disorders (Cluster A, B and C). Narcissistic Personality Disorder is part of cluster B, which is characterized by dramatic, emotional or impulsive behavior (APA, 2013). Cluster B also includes Borderline, Histrionic and Antisocial personality disorders. According to the DSM 5 (2013), Narcissistic Personality Disorder can be defined as a pervasive pattern of grandiosity, (in fantasy or behavior), need for admiration, and lack of empathy. It begins by early adulthood and is present in a variety of contexts.

The diagnosis for Narcissistic Personality Disorder is indicated by five of the nine criterion. The first criterion is having a grandiose sense of self-importance (e.g. exaggerates achievements and talents, expects to be recognized as superior without commensurate achievements). This is often a trademark of a narcissist, where they believe they are more beautiful, powerful and stronger than those around them. One example is exaggerating achievements; individuals can say they brought a company to great success, when it is not entirely true; they often take a small truth and exaggerate it out of proportion.

The next criterion is preoccupation with fantasies of unlimited success, power, brilliance, beauty, or ideal love (APA, 2013). Individuals with this criterion believe they have all of these things in themselves. This is one of the key aspects that propels them through difficult situations which is having a sense of being, or a strong sense of who they are. The third criterion is the person believes he or she is "special" and unique and can only be understood by, or should associate with, other special or high-status people or institutions (APA, 2013). This can be seen when individuals have a superior sense of self and see others as lower than they are.

The fourth criterion is need for excessive admiration (APA, 2013). Individuals with this disorder seem to need constant praise all the time. The fifth criterion is having a sense of entitlement (APA, 2013). They need a feeling or a sense they are powerful. They often seek compliments to make themselves have an inflated sense of self. The sixth criterion is that they are interpersonally exploitative (APA, 2013). An example of this is they take advantage of others to achieve their own ends.

The seventh criterion states individuals lack empathy: They are unwilling to recognize or identify with the feelings and needs of others (APA, 2013). With this criterion, they are seen as unable to associate or understand feelings of other people, but have an expectation for others to recognize their feelings. These individuals are also seen as having trouble maintaining relationships. The eighth criterion states people with NPD can be envious of others or believe that others are envious of them (APA, 2013). This can be related to their appearance, intelligence, career, etc. The ninth criterion says NPD individuals show arrogant, haughty behaviors or attitudes (APA, 2013). This can be seen as unrealistic goals. They believe they have the power or intelligence to change the goals when it is not practical to situations. These individuals can be seen as deeply insecure to the people in their lives.

There are other diagnostic features when considering Narcissistic Personality Disorder. Of those diagnosed with this disorder, 50-75% are male (APA, 2013). Approximately 18 percent of males and six percent of females have narcissistic traits. (McLean, 2007). The prevalence of NPD in the clinical population ranges from 2 to 16 percent and exists in the general population at a rate of less than one percent (Widiger & Mullins, 2003). There are certain features of this disorder that can be confused with other disorders or traits because they have certain characteristics in common. One example is mania or hypomania (APA, 2013). Grandiosity is often associated as part of Bipolar disorder as part of a manic episode, however the association with mood changes help distinguish the episodes from the disorder. In Obsessive Compulsive Personality disorder, individuals have a commitment to perfectionism and believe others can't do things as well (APA, 2013). In contrast, Narcissistic personality disorder patients are most likely to believe that they have achieved perfection.

Social Psychology Theory

Social Psychology can be defined as the scientific study of the effects of social and cognitive processes in the way individuals perceive, influence and relate to others (Smith, Mackie & Claypool, 2014). Social behavior is thought about in our everyday lives. For example, our decisions to make good friends, make decisions, raise children and live in peace rather than conflict are all ways we think about social behavior. (Smith, Mackie & Claypool, 2014). The presence of other people and what they pass on to us, and our feelings about the groups we belong to influence us in our social processes. Perceptions, memories and emotions are an influence on our cognitive processes (Smith, Mackie & Claypool, 2014). The people who surround us affect the way we behave in society. A classroom presentation or a job interview or a conversation with a boss all have to do with others observing us and interacting with us, which influences our thoughts, feelings and behavior.

Social psychology focuses on the way we relate to others through social and cognitive processes. It helps us to comprehend why individuals behave the way they do in situations and helps solves social problems. For example, examining divorce as an outcome of social and cognitive processes is a way in which social psychology studies conflict in marriages. Using a social psychology perspective, researchers explore how couples interpret events that put the relationship under stress and examine what alternatives to the relationship they think they have (Smith, Mackie & Claypool, 2014).

Research by Smith, Mackie and Claypool (2014) states that there are eight basic principles of social behavior. The first principle states people construct their own reality (Smith, Mackie & Claypool, 2014). This principle demonstrates our thoughts and actions are responsible for the perception of our own reality. The next principle says social environments pervasively influence people (Smith, Mackie & Claypool, 2014). In other words, the environment an individual lives in exerts a lasting influence over a wide range of behaviors.

The next three principles are considered "motivational principles." The first motivational principle asserts that people strive for mastery (Smith, Mackie & Claypool, 2014). This principle claims people seek to understand and predict events in the social world in order to obtain rewards. The next motivational principle conveys people seek connectedness (Smith, Mackie & Claypool, 2014). This mentions people seek support, liking, and acceptance from the people and groups they care about and value. The third motivational principle is that people value "me and mine" (Smith, Mackie & Claypool, 2014). It suggests people desire to see themselves, people and groups connected to themselves, in a positive light.

Finally the last three principles are called processing principles (Smith, Mackie & Claypool, 2014). These operate together to help guide our cognitive efforts in understanding ourselves and others. These are Conservatism, Accessibility, and Superficiality versus Depth (Smith, Mackie & Claypool, 2014). Conservatism suggests that people are slow to change (Smith, Mackie & Claypool, 2014). Accessibility states information that is accessible has large effects (Smith, Mackie & Claypool, 2014). These have the most impact on our thoughts, feelings, and behaviors. The final principle is superficiality versus depth (Smith, Mackie & Claypool, 2014). This principle implies that people put in little effort when dealing with information. When it is superficially handled, they can be convinced to process the information at greater depth.

Social psychology utilizes a variety of theories for cognitive and social phenomena. One of which is Social Comparison Theory, which explains how individuals evaluate their own opinions and abilities by comparing themselves to others (Festinger, 1957). This is seen to reduce uncertainty in specific domains, and the individual learns how to define the self. This can be seen in relation to NPD because individuals often try to achieve a higher sense of self compared to others around them.

Festinger (1957) developed three hypotheses to better understand social comparison processes. The first hypothesis states every human has a drive within themselves to evaluate his or her opinions and abilities (Festinger, 1957). The similarities between opinions and abilities is that they are both affected by our behavior. When an individual is punished for a certain opinion or ability, it can alter the way one behaves. For example, when examining one's ability in a sport, one can view themselves as being a good sports player. However, the opinions of others on how they play that certain sport can influence their thoughts and ultimately their performance as well.

This hypothesis also supports the idea of comparing one's performance to another individual who did significantly better. An example that Festinger (1957) uses is running around a track and timing oneself. The individual is most likely going to compare his or her time to other individuals who ran the same track. The second hypothesis states that people evaluate their opinions and abilities based on the opinions and abilities of others (Festinger, 1957).

A question Festinger (1957) raised was how does one determine if a political candidate is better than another? This question was implied to demonstrate when an individual listens to the opinions of others, it has an effect of altering their own thoughts, based on how much influence the person has on them. When examining "ability" to support this hypothesis, Festinger raises a question of how one determines how intelligent one is. For example, when looking at a test which measures for intelligence, one cannot come to a conclusion they are smart without comparing their scores to other individuals who took the test.

The last hypothesis asserts the tendency to compare oneself with some other specific person decreases as the difference between his opinion or ability and one's decreases (Festinger, 1957). In other words, one does not compare their own opinions and abilities to those that are too divergent from them. For example, an individual who is a novice at playing an instrument does not compare themselves to individuals who have mastered the talent. Festinger (1957) also states this can be applied to a person's opinion. In today's society, there are debates on social issues such as being gay, abortion etc. An individual who has a stance on a specific social issue will not evaluate his opinions based on those who take a stance on the opposite side.

Narcissistic Personality Disorder through a Social Psychology Lens

Social psychology looks at the conditions in which certain behaviors may occur. In our society, the ways an individual is motivated have changed through Western culture over the years. For example, competition has risen since the Great Recession in 2008 (Porter, 2011). Competition can be seen in biology, ecology, and sociology in contests between organisms, animals, individuals and groups. They can be in competition for a variety of things such as mates, location, resources, prestige, or power.

There are different arguments which can critique the diagnosis of Narcissistic Personality Disorder. One example is how individuals are motivated by the Economic Recession. In a Social Psychology lens, a question can be raised on what motivates individuals to behave or act the way they do. In a society that promotes competition, it is normal for one to feel that need to be better equipped than others. One example has to do with executives or CEO's of companies. Peer pressure, pride, and the desire to look good in the community spur executives to outdo one another (Potter, 2011), a characteristic of NPD.

The first criterion for Narcissistic Personality Disorder states that the individual has a grandiose sense of self-importance (exaggerating one's achievements and talents). In today's society, which is fueled by competition, one can argue that this is not a dysfunctional act. For example, in the Western culture, the quote "to achieve the American dream" has been repeated and even taught to us. The "American Dream" can be seen as an opportunity for prosperity and success. As stated before, perceptions, memories and emotions are an influence on our cognitive processes (Smith, Mackie & Claypool, 2014). If the "American Dream" was introduced in childhood or by peers at an older age, this can shape how we behave and interact with other individuals.

The second criterion is preoccupation with fantasies of unlimited success, power, brilliance, beauty, or ideal love (APA, 2013). Advertisements and social media play a role in defining who we should be as individuals. The majority of individuals who show up on the television screens, advertisements, movies, are celebrities. These influences are hard to miss. They are often portrayed in order to achieve the highest standards of beauty and wealth.

One of the principles mentioned in social psychology stated social environments pervasively influence people (Smith, Mackie & Claypool, 2014). In an environment where media is a powerful influence on what it looks like to achieve unlimited success, individuals can be bound to mimic that image and take it on themselves and portray it to other people. Individuals can also purchase items that can make them feel wealthy or beautiful; however it does not reflect what they are truly feeling on the inside.

The third criterion is the person believes he or she is "special" and unique and can only be understood by, or should associate with, other special or high-status people or institutions (APA, 2013). A motivational principle in social psychology states people seek support, liking, and acceptance from the people and groups they care about and value (Smith, Mackie & Claypool, 2014). Individuals seek connection and acceptance, whether this is from a boss, peer, family member or partner. This criterion can be critiqued recognizing that individuals have different motivations and drives behind their behavior.

The fourth criterion is excessive admiration (APA, 2013). Individuals who are often diagnosed with Narcissistic Personality Disorder need constant admiration all the time. The motivational principle "me and mine" suggests people desire to see themselves, and other people and groups connected to themselves, in a positive light (Smith, Mackie & Claypool, 2014).

Normative social influence states we comply in order to fuel our need to be liked or belong. Praise and admiration is often what makes us feel good and reinforces behaviors we do. With an individual diagnosed with Narcissistic personality disorder, feelings of emptiness can explain their need for so much admiration and praise from people, which can come off differently to people they interact with.

The fifth criterion is having a sense of entitlement (APA, 2013). Patients with this criterion are often seen to have unreasonable expectations and favorable treatment in compliance with his or her expectations. In today's society, technology has made everything easier for us. With the new trend in advanced phones, individuals have to do little to no work to get something accomplished. For example, if you are out on the road and you need to look up directions or information on your e-mail, it is all in your smart phone at your convenience. In this generation, children are often seen with phones or tablets. These devices are also used as reinforcement to praise a behavior. There are more inventions created every day that makes our lives easier and as a result can make us feel more entitled or impatient with others. Social Psychology states that we are prone to social influence even when no other people are present, such as watching television or following internalized cultural norms.

The sixth criterion states they are interpersonally exploitative (taking advantage of others for his or her gains (APA, 2013). Going back to the discussion of competition, when we feel we lack something we are often in the mode of survival. For an individual who is used to getting what he or she wants, he or she can take advantage of people who will easily comply to give them what they need. The accessibility principle states information that is accessible has large effects (Smith, Mackie & Claypool, 2014). Individuals who comply with getting taking advantage of are often the ones who reinforce the behavior of the narcissist; because the individual has a technique of achieving what they want, it shapes their behavior and how they will treat others.

I will critique the last three criteria using Social Comparison, a theory that is used under Social Psychology. Social Comparison Theory explains how individuals evaluate their own opinions and abilities by comparing themselves to others in order to reduce uncertainty in these domains, and learn how to define the self (Festinger 1957). The seventh criterion states the individual lacks empathy and is unwilling to identify the needs and feelings of others. Festinger (1957) states in one of his hypotheses that one does not compare his or her own opinions and abilities to those that are to divergent from them. Individuals who are labeled as Narcissistic may not identify with the feelings of the other person they are interacting with because they do not share the same views. They often may be seen as unempathetic because they are constantly making self-evaluations based on others they idealize and value. Being vulnerable can be difficult in a society that is highly individualistic.

The eighth criterion asserts individuals with narcissistic personality disorder are often envious of others or others are envious of them (APA, 2013). Festinger (1957) mentions in his first and second hypothesis that every human has a drive within themselves to evaluate his or her opinions and abilities and also evaluate others as well. This goes back to the discussion of competition and how everyone is trying to reach the highest standards in today's society.

Envy is one of the normal feelings human beings feel, as well as anger, sadness, etc. This is not just seen with individuals who are labeled as Narcissistic. We live in a society where things are not measurable without comparing ourselves to other people (Porter, 2011). As mentioned earlier, taking a test to measure intelligence will not tell you how smart you are unless compared to scores of other individuals who took the same test.

The last criterion mentioned in this diagnosis is arrogant, haughty behaviors or attitudes (APA, 2013). Festinger's (1957) third hypothesis states one does not compare their own opinions and abilities to those that are different. Individuals can be passionate about certain topics when having a conversation and can often come off as arrogant or haughty to people who have the opposite view. Most of us have the social skills and impulse control to keep our envy and social comparisons quiet but our true feelings may come out in subtle ways.

Narcissistic Injury is a term used to describe an individual's feelings of self-worth based on a variety of factors, such as employment, relationships, education, etc. When an individual fails in any of these domains, they experience narcissistic injury, in which they feel worthless because all of their self-worth is on achieving or obtaining a certain goal. Clinician's objectives are to help these individuals learn how to have a healthy sense of self without relying on another object to define their worth.

Conclusion

Narcissistic Personality Disorder is distinctive in its criteria; however when looking at it through a Social Psychology lens one can question if it should be considered a mental disorder. Individuals with this disorder may be a product of society, which can influence our thoughts, feelings and behaviors. Social Comparison Theory also states we are looking at our own worth based on how we compare ourselves against others. Based on the lens of Social Psychology, it is my understanding that individuals labeled as "Narcissistic" do not have a disorder but are following the internalized cultural norms of an individualistic society which influences their behavior.

References

Diagnostic and statistical manual of mental disorders: DSM 5. (5th ed.). (2013). Washington, D.C.: American Psychiatric Association.

Festinger, L. (1954). A Theory Of Social Comparison Processes. *Human Relations*, 117-140.

McLean J. *Psychotherapy with a Narcissistic Patient Using Kohut's Self Psychology Model.* Psychiatry. 2007;4:40–47

Huggins, R., & Porter, M. (2011). Clusters and the New Economics of Competition. In *Competition, competitive advantage, and clusters the ideas of Michael Porter.* Oxford: Oxford University Press.

Smith, E., Mackie, D., & Claypool, H. (2014). *Social psychology* (Fourth ed., p. 740). Psychology Press.

Widiger TA, Mullins S. *Personality disorders.* In: Tasman A, Kay J, Lieberman J (eds). Psychiatry, Second Edition. West Sussex, England: John Wiley & Sons, Ltd., 2003:1623–5.

Schizophrenia: Transcending the Bio-Social Divide to a Feminist Embodiment

Renee Dieschbourg

Introduction

"A misfit occurs when world fails flesh in the environment one encounters—whether it is a flight of stairs, a boardroom full of misogynists, an illness or injury, a whites-only country club, subzero temperatures, or a natural disaster" (Garland-Thomson, 2011, p. 600). This concept of "fitting" encompasses an intersection of identities *intra-acting* with the environment. One could argue in our patriarchal, capitalist, American society that an educated white, able-bodied/minded, heterosexual, middle to upper class male is most protected, supported, and thus sustained by his environment. So what happens to those who do not embody the above identities?

In Western culture mental illness is the most stigmatized and marginalized type of disability, and I would argue schizophrenia is portrayed as the "craziest" hallmark diagnosis. The effects of being labeled with a diagnosis or "othered," has real consequences, which only become further complicated when additional oppressed identities intersect. In this paper, I will (re)examine the diagnostic category schizophrenia with a feminist standpoint using disability theory, intersectionality, and concepts of embodiment from material feminism. The purpose of using this framework and theory is for two critical reasons: to reveal how systems of oppression are interconnected and to value subjugated knowledges.

Constructing Pathology

Feminist researchers, theorists, and scholars are invested in uncovering women's and other marginalized groups'--such as people of color, queer, poor, and people who have a disability or illness-- lived experiences. The hope is to carve a space for these groups that experience systemic oppression to have voice, thus allowing them to define their own reality from a non-dominant standpoint. Also, feminists work to expose how dominant groups are privileged and why normalizing or universalizing privileged groups' experience is problematic. People who do not "fit" society's rigid constructs of "normal" have been feared and punished; this is especially true for women because traditional notions of femininity are so limiting: piety, purity, submissiveness, and domesticity.

Throughout history stories of madness, hysteria, insanity, psychosis, or deviance have been documented, each construed, constructed, and conceptualized relative to time period, place, and author's standpoint. The late 1600's mark an unforgettable time in American History of "othering;" witch-hunting was a movement that accused vulnerable people, or those who threatened the power structure of witchcraft or hysteria, usually women, in order to maintain social control. They were thought to be possessed by the devil or to embody mystical powers, and many were hanged. Reed (2007) suggests that the Puritans' focus on witchcraft is situated in a social context with a deeper underlying meaning: "a conception of masculinity tied to the Puritan image of God's absent efficacy, and a conception of femininity tied to the Puritan image of the Devil's constant, present temptations of body and soul" (p. 210-11). By the 1800's witch-hunting had phased out and the development of Psychology began as a new means of controlling the "other."

In the late 1800's Emil Kraepelin began expanding on Morel's 1852 construct of démence précoce or dementia praecox. Through studying and observing the behaviors of asylum inmates in Europe, Kraepelin began to group clusters of behavior. Kraepelin's writings, dating back to 1896, are considered to mark the beginning of an attempt to define what is now referred to as schizophrenia (Boyle, 1990). Influenced greatly by Kraepelin, Eugen Bleuler, published *Dementia Praecox or the Group of Schizophrenias* in 1911. Bleuler became more influential in the United States and coined the term *schizophrenia*, stating that dementia praecox "only designates the disease, not the diseased" (Boyle, 1990, p. 60). This begins to further pathologize those diagnosed; dementia praecox is the disease, and the person diagnosed is a schizophrenic.

Wide variations in behavioral criteria of this diagnostic category can be seen beginning with Kraepelin, declaring up to eleven major sub groups within the construct of dementia praecox. This stretch continues with the Diagnostic and Statistical Manual of Mental Disorders (DSM 5), now referring to schizophrenia as a spectrum (Boyle, 1990). According to the DSM 5, the diagnostic criteria for schizophrenia includes: have two or more of the following consistently for one month: delusions, hallucinations, disorganized speech, grossly disorganized or catatonic behavior, and negative symptoms and that these symptoms affect one's level of functioning in one or more major areas such as work, interpersonal relations, self-care (American Psychiatric Association, 2013, p. 99). Read (2010, p. 10) states that schizophrenia "…is also disjunctive, meaning that one person can receive the diagnosis without having anything in common with another person with the same diagnosis" (Bentall, 2003,2009; Read, 2004b). This wide variance in the way symptoms manifest in different people has lead some researchers to reject the term altogether and focus on specific phenomenon like hallucinations and delusions (Read, 2010).

Psychiatry and Psychology, even in more contemporary times, are structured around pathologizing the person; this is especially true with the diagnosis of schizophrenia. The target group to pathologize tends to be the vulnerable or the "other," those that do not "fit" with society's dominate norms. Mothers historically have been blamed for causing schizophrenia in their children, which lead to the term *schizophrenogenic mother*, yet another example of a constructed label for the person, not the illness (Gabbard, 2005, p.191). Thomas Szasz (1974) states that "therapeutic interventions have two faces: one is to heal the sick, the other is to control the wicked" (p. 69).

In addition to women as witches and as schizophreniogenic mothers, people living in poverty became another misfit population to target. In 1939, Faris and Dunham produced a ground breaking study revealing, for the first time, the disproportionate rate of diagnosing based on neighborhood, "People in the poorest areas of Chicago were seven times more likely to be diagnosed 'schizophrenic' than those in the richest parts" (Read, 2010, p. 11). Poverty is clearly a relentless stresser, or constant trauma to one's mental health, however it is important to notice the power dynamic between the wealthy and poor and consider mass diagnosing in poor neighborhoods as a form of social control. "By 1977 there were nine studies showing that more severe diagnoses are applied to poorer people than wealthier people with the same symptoms" (Abramowitz & Dokecki, as cited in Read, 2010, p. 13). This is a critical piece to demystifying DSM 5 diagnoses because it allows us to start asking questions. Who decides what is normal and who has the power in society to regulate and dictate normalcy?

The pharmaceutical industry plays a huge part in constructing pathology in order to grow business. Drug companies fund about half of all mental health websites, and within those websites information is rooted in a biological understanding, which strengthens the justification of medication as treatment (Read, 2010). A study done by Cosgrove, Krimsky, Vijayaraghavan, & Schneider (2006), revealed that "Of the 170 DSM panel members 95 (56%) had one or more financial associations with companies in the pharmaceutical industry" (p. 154). This is an extreme conflict of interest; pharmaceutical companies with ridiculous amounts of wealth and power decide what is "normal." This greatly influences the construction and treatment of mental illness; medication is the primary, universal treatment for schizophrenia. This is troublesome in two ways: it causes some practitioners to throw out the bio-psycho-social model by only focusing on the biological, and it calls into question whether medication is being overused with this diagnosis (Read, 2010). Intersectionality is one way to better understand how systems influence the diagnostic construct of schizophrenia.

Intersectionality

Kimberlé Crenshaw established the concept of intersectionality to explain how identities and oppressive systems intersect and are inseparable in peoples' lived experience. The Combahee River Collective, a black feminist group, further articulate intersectionality, "We also often find it difficult to separate race from class from sex oppression because in our lives they are most often experienced simultaneously" (1981, p. 213). Disability is another marginalized identity that intersects with other systems; the stigma of being labeled schizophrenic further marginalizes women, people of color, queer, poor, and other discriminated groups.

Johnstone (2000, p. 238) explicitly states how race and class intersect with being diagnosed with schizophrenia, "Working class patients are, like black and ethnic minority patients, more likely to be prescribed physical treatments such as drugs and ECT, to spend longer periods in hospital regardless of diagnosis, and to be readmitted, and, correspondingly less likely to be referred for the more 'attractive' treatments such as psychotherapy or group therapy" (Read, 2010, p. 14). People who have been marginalized or "othered" have a valuable perspective and subjugated knowledge of the world, but in dominant systems only privileged/ "normal" voices are heard, which is why exposing different standpoints is extremely important, especially from those diagnosed.

Feminist Standpoint

Black feminists have been exceptionally active in revealing whose knowledge and voices are privileged and represented and why "others" are devalued, forgotten, and invisible. Patricia Hill Collins explains, "Suppressing the knowledge produced by any oppressed group makes it easier for dominate groups to rule because the seeming absence of an independent consciousness in the oppressed can be taken to mean that subordinate groups, willingly collaborate in their own victimization" (1990, p. 5). In order to understand the diagnostic construction of schizophrenia, it is critical to recognize whose standpoint has historically dominated the fields of Psychology and Psychiatry. White, educated men, the dominant, privileged group in society, have widely been the ones researching, creating theories, and writing about the experiences that those "other," subordinate people have. "Feminist standpoint epistemology focuses on the scientific and epistemological importance of the gap between the understanding of the world available if one starts from lives of people in the exploited, oppressed, and dominated groups and the understanding provided by the dominant conceptual schemes" (Harding, 1993, p. 147).

Moreover, Black feminist scholars point out how the dominant standpoint and language distorts oppressed groups' ideas, because processing an idea through a privileged paradigm is a disconnect, which unavoidably misrepresents the meaning (Collins, 1990). Feminist standpoint theorists believe that knowing and knowledge are not relative, but rather partial; the oppressed/marginalized are thought to have a more authentic and informed standpoint. This is called "double consciousness;" people located in oppressed groups must function and understand dominant discourse, while also understanding their own lived experience, the experience as the marginalized "other." Biological illness and social phenomena are situated in a time and place; therefore I find disability theory helpful in deepening our understanding of schizophrenia.

Disability Theory
Deconstructing the "Other"

Wendell articulates the experience of illness creatively: "Socially constructed from a biological reality" (1989, p. 107). It is not about determining if schizophrenia is biological/genetic through a reductionist medical model versus a social construct, but rather acknowledging that bodies and the environment *intra-act* and peoples' lived experiences of illness are quite real. Disability theory repositions individual fault or pathology, by turning the focus on the culture, environment, and larger society. Garland-Thomas (2011) speaks about this interconnectedness, "Disability oppression in this view emanates from prejudicial attitudes that are given form in the world through architectural barriers, exclusionary institutions and the unequal distribution and access to resources" (p. 591). Disability activists force American society to realize how able-bodied and minded people are valued, therefore institutions and systems support and sustain this privileged group.

Garland-Thomas goes on to state, "Our conventional response to disability is to change the person through medical technology, rather than changing the environment to accommodate the widest possible range of human form and function" (2011, p. 603). Allowing marginalized voices to be heard and to understand that people with diagnoses have important and valid knowledge is at the foundation of both feminist and disability theory. Marie, a woman diagnosed with schizophrenia, speaks out, "the focus on the sick model as opposed to the able models is creating a lot of self-perceptions that the person is sick." Marie goes on to talk about her healing (re)concepualization of schizophrenia stating, "…a disability is uniqueness, and you're not broken, you do not need to be fixed" (Schneider, 2007, p. 136). A study cited below demonstrates how biology and environment overlap, thus (re)aligning the focus on Person-In-Environment or "fit."

Kendler and Eaves (1986) postulated that genes control the degree to which an individual is sensitive to the environment's predisposing, risk-increasing aspects versus its risk-reducing and protective aspects" (Gabbard, 2005, p. 192). A Finnish study looked at a group of adoptees with biological mothers diagnosed with schizophrenia compared to a group of adoptees with no increased genetic risk. When the group of adoptees with higher genetic risk were placed in a home with parents' who struggle to effectively communicate, also known as communication deviance, they exhibited higher rates of thought disorder compared to the group with no increased genetic risk. Gabbard (2005) then concludes, "In this conceptual model, emphasis is placed on the "fit" between child and family" (p.192). The interaction of genetics and environment is a perfect example of the fluidity and porosity that disability theory advocates; the two are not separate.

"When we make people 'other,' we group them together as the objects of *our* experience instead of regarding them as fellow *subjects* of experiences with whom we might identify" (Wendell, 1989, p. 116). People with schizophrenia have a different style of communication, or perhaps society's dominant way of communicating is simply unfitting to them. Labeling and "othering" people perpetuates an atrocious system that justifies controlling, criminalizing, and stripping away one's human rights. A relatively newer area within feminist discourse is Material feminism, which compliments disability theory by transcending either/or thinking and breaking through restrictive dichotomies. Conceptualizing schizophrenia cannot be done with a binary framework; the real or material and socially constructed phenomenon are interconnected. There is fluidity in illness; thinking about schizophrenia and all mental illness must start with each individual's lived experience. Bost (2008) talks about the real risk normal/abnormal dichotomies have, "Vilifying the sick as the constitutive outside of a healthy society involves not paying attention to the content of sickness, drawing boundaries rather than healing, sorting us out to keep the queer and ill away for the mythlogized healthy family" (p. 347).

Feminist Materiality

Historically, male philosophers dichotomized the sexes, creating mind-body dualism; thus attributing the valued mind to man and the devalued body to woman (Nicki, 2001). This misogynistic claim was purposely linked to deeming women as inferior to men. Feminists have fought long and hard to reject woman's association with the body and the corporeal. However, in more recent times, feminists have come to understand that abandoning the body is also problematic.

Material feminism embraces mind and body connection, as do disability activists. Bost says, "Studying the fluid matter of illness itself, in defiance of the medical and cultural discourses that demonize such matter, problematizes the identities and distinctions upon which boundaries rest" (2008, p. 348). Materiality is the notion of messy interconnectedness; the overlap, the entanglement, and the exchange of the body or corporeal and the environment or socially constructed phenomenon must be acknowledged. Butler in *Bodies that Matter*, clearly states, "...surely bodies live and eat; eat and sleep; feel pain and pleasure; endure illness and violence; and these facts ... cannot be dismissed as mere construction" (1993, p. ix). It becomes apparent that the mind/body, biological/social dichotomies and either/or thinking are limiting and stigmatizing.

Embodiment

Once this new materialism perspective is understood, embodiment is the praxis, the manner of *being in the world*. Eleanor Longdon and Elyn Saks, two successful women who have given Ted Talks about living with a diagnosis of schizophrenia claim that making meaning out of the illness can be empowering. Gabbard (2005) reinforces this understanding that a person's behaviors are a reaction/response to an unsustainable/threatening environment; "Psychotic symptoms have meaning. Grandiose delusions or hallucinations, for example often immediately follow an insult to a schizophrenic patient's self-esteem" (Garfield 1985: Garfield et al. 1987, as cited in Gabbard, 2005, p. 193). Listening to the voice's tone and content and recognizing when one experiences delusions or hallucinations can reveal repressed emotions from trauma. Making meaning out of one's marginalized experience and embracing difference allows one to reclaim one's experience and shed pathologizing identities

Eleanor Longdon advocates for a shift in language, renaming her experience "hearing voices," which stemmed from an international empowerment movement called the Hearing Voices Network. The Hearing Voices Network believes that individuals are unique and have the right to self-determination in order to work through their psychotic episodes. As social work practitioners, shifting our language and perceptions will help us better understand our clients and limit "othering." Szasz (1974) reframes difference as follows, "persons said to be schizophrenic often exhibit a certain unconventional manner of using language" (p. 32). This empowerment approach is inherently feminist and also compliments Disability theory. Embodying difference is interwoven into material feminism and disability theory. Viewing schizophrenia from a feminist and disability standpoint allows one to see the strengths, uniqueness, and humanness of the person behind the diagnosis.

Closing

Thomas Szasz once famously said, "Insanity is the only sane reaction to an insane society" (Szasz, 1974). As social work practitioners, it is crucial that we understand how intersecting power systems and marginalized identities influence our clients' mental health. If we ask the right questions, it will become clear how body and environment *intra act*. Certainly people of privileged groups also experience mental illness, however the intersection of marginalized identities simply makes one more vulnerable due to the added stress of daily life and lack of resources. I would like to end with Garland-Thomson's quote, "When we fit harmoniously and properly into the world, we forget the truth of contingency because the world sustains us" (2011, p. 597).

References

American Psychiatric Association. (2013). *Diagnostic and statistical manual of mental disorders* (5th ed.). Washington, DC: American Psychiatric Association.

Bost, S. (2008). From race/sex/etc. to glucose, feeding tube, and mourning: The shifting mater of Chicana feminism. In Alaimo, S. & Hekman, S. (Eds.), *Material Feminisms* (340-372). Bloomington: Indiana University Press.

Boyle, M. (1990). *Schizophrenia: A scientific delusion?.* New York: Routledge.

Butler, J. (1993). Bodies that Matter: On the Discursive Limits of Sex, London: Routledge.

Collins, P. (1990). Black Feminist Thought: Knowledges, Conscioiusness, an the Politics of Empowerment. New York: Routledge.

Combahee River Collective (1981). A black feminist statement. In Moraga, C. & Anzaldúa, G. (Eds.). *This Bridge Called My Back: Writings by Radical Women of Color.* (210-218). New York: Kitchen Table: Women of Color Press.

Cosgrove, L., Krimsky, S., Vijayaraghavan, M. Schneider, L. (2006). Financial ties between DSM-IV panel members and the pharmaceutical industry. Psychotherapy and psychosomatics, 75, 154-160.

Gabbard, G. (2005). *Psychodynamic psychiatry in clinical practice* (5th ed). Washington, D.C.: American Psychiatric Press.

Garland-Thomson, R. (2011). Misfits: A feminist materialist disability concept. *Hypatia 26*(3),   591-609.

Harding, S. (1993). Reinventing ourselves as other: More new agents of history and knowledge. In Kauffman, L. (Ed) *American Feminist Thought at Century's End: A Reader.* (140-164). Cambridge: Blackwell Publishers.

Nicki, A. (2001). The abused mind: Feminist theory, psychiatric disability, and trauma. *Hypatia 16*(4), 80-104.

Read, J. (2010). Can poverty drive you mad? 'Schizophrenia', Socio-Economic Status and the   case for primary prevention. *New Zealand Journal of Psychology 39*(2), 7-19.

Reed, I. (2007). Why Salem made sense: Culture, gender, and the Puritan persecution of   witchcraft. *Cultural Sociology 1*(2), 209-234.

Schnieder, B. (2007). Constructing a schizophrenic identity. In Raoul, V., Canam, C., Henderson, A., and Paterson, C. (Eds.). *Unfitting Stories: Narrative Approaches to   Disease, Disability, and Trauma.* (129-137). Waterloo: Wilfrid Laurier University Press.

Szasz, T. (1974). *The Myth of Mental Illness: Foundations of a Theory of Personal Conduct.* (Revised Ed.). New York: Harper & Row, Publishers.

Wendell, S. (1989). Toward of feminist theory of disability. *Hypatia 4*(2), 104-124.

Social Learning Theory and Addictive Disorders

Paige Salk

Addiction is something that I have always found very interesting and something that I have wanted to learn more about. Addiction comes in many different forms. For this paper I will be focusing on substance abuse and alcohol addiction.

As a person studying for her masters in social work, I have learned about several theories and the way each illustrates why things are the way that they are. Personally, I have always found social learning theory to be the most interesting. I believe that individuals can learn by being around others and by watching the way that people act and observing what they do. I believe a person's environment and the people they surround themselves with greatly increase one's risk for substance abuse disorders. For this paper, I will use the social learning theory as a way to get a different type of perspective on addiction.

Before we look at the social learning theory and how it views addiction, lets first get a better understanding of substance use addiction. According to the DSM V, "the essential feature of a substance use disorder is a cluster of cognitive, behavioral, and physiological symptoms indicating that the individual continues using the substance despite significant substance-related problems" (American Psychiatric Association, 2013, p. 483).

In order to better understand the different categories of addiction, they are broken down into different criteria. "Impaired control over substance abuse" is criteria one through four (American Psychiatric Association, 2013, p. 483). Examples of this criterion include that, "the individual may take the substance in larger amounts or over a longer period than was originally intended, the individual may report multiple unsuccessful efforts to decrease or discontinue use, the individual may spend a great deal of time using the substance and recovering from its effects and all of the individuals' activities revolve around the substance" (American Psychiatric Association, 2013,p. 483).

The second grouping of the criteria for substance use disorders is social impairment. This falls under criteria five through seven (American Psychiatric Association, 2013). Examples of social impairment include "individual may continue substance use despite having persistent or recurrent social or interpersonal problems caused or exacerbated by the effects of the substance" (Ameri- (American Psychiatric Association, 2013, p. 483).

The third grouping, criteria eight through nine are risky use of the substance (American Psychiatric Association, 2013). Criterion eight refers to an individual continuing use of the substance in situations that are considered physically hazardous. Criterion nine refers to the individual recognizing that they have a problem that has been caused by use of the substance, but continuing to use the substance anyways. "The key issue in evaluating this criterion is not the existence of the problem, but rather the individual's failure to abstain from using the substance despite the difficulty it is causing" (American Psychiatric Association, 2013, p. 483).

The last grouping, ten through eleven, is pharmacological criteria. Criterion ten refers to tolerance, which "is signaled by requiring a markedly increased dose of the substance to achieve the desired effect or a markedly reduced effect when the usual dose is consumed" (American Psychiatric Association, 2013, p. 484). For example, the more often that a person consumes alcohol, the more that person's tolerance will increase. When a person first starts drinking or using drugs, it might only take them a small amount to feel the desired effect. However, the more frequently the person uses alcohol or drugs, the larger the amount they will have to use in order to feel that desired effect.

Criterion 11 refers to withdrawal, which "is a syndrome that occurs when blood or tissue concentrations of a substance decline in an individual who had maintained prolonged heavy use of the substance" (American Psychiatric Association, 2013, p. 484). This means that when a person has been using a substance and their body has gotten used to that particular substance being in their system, when the person stops taking that substance, their body starts to go through withdrawal. Withdrawal symptoms vary greatly depending on the type of drug the person is using. When withdrawal symptoms start to occur, regardless of the type of drug, the person is likely to continue using the drug to relieve the withdrawal symptoms.

Now that we have a better understanding of how substance use disorders are classified, let's try to better understand the social learning theory. In order to understand social learning theory, it is important to understand that "social learning theory is composed of four major components: differential association, definitions, imitation, and differential reinforcement" (Peralta & Steele, 2010, p. 866).

According to differential association, bad behavior or criminal behavior is learned through social interactions, such as friendships or relationships. "Learning criminal and/or deviant behavior includes learning the techniques, motives, rationalizations, and attitudes needed for committing the violation. Learning bad or criminal behavior "follows the same processes associated with learning other behaviors" (Peralta & Steele, 2010, p. 866).

People learn what is good versus bad and right versus wrong through interactions with significant groups in their lives. Social learning theory asserts that "the more individuals define the behavior positively or as justified, the more likely they will engage in it" (Akers, Greca, Cochran & Sellers, 1988, p. 626). For example, if a person associates with others who drink alcohol frequently and enjoy drinking alcohol, the more likely that person is to drink alcohol as well. This is because the individual is only seeing positive alcohol associations and sees the outcome of drinking alcohol as being something positive. In easier terms, "how we think affects the way we behave; but how we behave also affects how we think" (Niaura, 2000), which is referred to as reciprocal determinism. The author goes on to explain the concept of efficacy expectations, which he describes when he writes that, "effortful behavior is thought to be mediated to a significant degree by the confidence we have in our ability to do something" (Niaura, 2000, p. 156). Social learning theory uses efficacy expectations in explaining effortful behaviors.

Social learning theory posits that a person can learn a behavior in many different situations. One main type of learning is through observation. If a person has watched others use drugs and seen that it was a pleasurable experience for them, that person is more likely to try that drug because they will associate the drug with seeing a positive and pleasurable outcome. On the other hand, if a person only has seen negative experiences with drugs, such as a parent becoming an addict and changing before their eyes in a negative way, or having their life revolve around using, they might associate drugs with being nonpleasurable and other negative outcomes, causing them to instead be less likely to use drugs. This specific social learning theory concept is one that brings up many questions. We see a lot of addiction that runs in families, where a child was born to a parent who is an addict and sees how it negatively affected both of their lives, yet the child turns to drug or alcohol use at an early age and may fall into the negative cycle. Although the child had seen the negative affects that drug use had on the parent and family, the child still chose to use.

I found an interesting study that was administered in several classes at a small university in the Midwest. The study was conducted to see the prevalence of nonmedical prescription drug use (NMPD) among college-aged students. The study defined nonmedical prescription drug use as "the use of prescription drugs for non-prescription purposes, specifically recreational use (which includes drug use in order to aid in studying, test taking, and getting high or buzzed)" (Peralta & Steele, 2010, p. 877). It was found that the average onset age of nonmedical prescription drug use was 18 years old and "a minority of students (39%) reported NMPD use at least once in their lifetime. About 31% reported NMPD use in the last year, 23.4% in the last three months, and 14.4% in the last thirty days. Also, 24% of those who had used prescription drugs non-medically reported that they used more than one prescription drug" (Peralta & Steele, 2010, p. 880). It was found that opiates were the most common drug used, followed by stimulants and then depressants. More specifically, it was found that Vicodin, Adderall and Ritalin were the three most common drugs that students reported using (Peralta & Steele, 2010, p. 880).

Nonmedical prescription drug use is a prevalent problem on college campuses today, and this study proved to be very relevant. It would be interesting to conduct this study among a variety of different universities in the Midwest, the West and the East coast and in the South and to compare the NMPD among these different college campuses.

The authors made an interesting point when they stated "most college students get their drugs from other students but fail to make the connection as to why certain individuals will choose to engage in drug use, while others refrain from this particular form of substance abuse" (Peralta & Steele, 2010, p. 877). Nonmedical prescription drugs are very easy to get on college campuses when many students are actually prescribed these drugs and are willing to give or sell their drugs to their peers. As discussed earlier, social learning theory explains that deviant behavior can be learned through interactions with peers. It is important to understand that "critical conditions for social learning theory to operate include exogenous (social concerns such as the impact of friends who use) as well as endogenous conditions (physiological or psychological precursors to substance use)" (Peralta & Steele, 2010, pg. 866). Since NMPD is so prevalent on college campuses already, if an individual's peers are using, the individual will be more likely to engage in the drug use as well.

Another interesting topic to look at, within substance abuse, is relapse. I found several different models that use social learning theory while explaining relapse. The first is called Marlatt's (1985) model of the relapse process. In this model, the person is considered to be in a high-risk situation and in order "to avoid relapse, the individual must execute a cognitive or behavioral coping response, which in turn produces an increase in self-efficacy" (Brandon, Herzog, Irvin, Gwaltney, 2004, p. 53). In turn, "if a coping response is not used, the individual experiences a decrease in self-efficacy and an increase in positive outcome expectancies for the initial use of the substance" (Brandon et al., 2004, p. 53). Positive outcome expectancies are also called cravings for the drug, and lead to the relapse.

This model makes a lot of sense. If a person who is a recovering addict is put into a situation where they are around drugs, in a place where they used to use those drugs, or with people who they associate with their drug use, they are in a high-risk situation. If recovering addicts do not have the proper tools or support system to help themselves through this situation, they are more likely to experience a relapse, compared to a person who does have the tools and coping skills to help them through this type of situation.

The second model that I found is by Niaura et al. (1988). This model also starts with a "contextual cue associated with drug use. This cue produces a series of responses, including urges to consume the drug, positive drug outcome expectancies and physiological activation" (Brandon et al., 2004, p. 54). These three responses "threaten the person's self-efficacy both directly and via diminished cognitive and behavioral coping. And low self-efficacy then increases the risk of relapse" (Brandon et al., 2004, p. 54).

The last model that I found is Cooper, Russel & George's (1988) alcohol abuse and dependence model. In this model, drinking to cope with stress is considered "the proximal predictor of heavy drinking" (Brandon et al., 2004, p. 54). Using drinking to cope is considered to be a poor coping mechanism and positive expectancies about what alcohol does to a person's mood. In this model, the individuals who are abusing alcohol do not have any effective coping skills because they are used to using their drinking as a means to cope with stressful situations in their lives. In this type of situation, when a person has quit drinking but they encounter a stressful event in their life, the first thought they have is to associate that stressful life event with alcohol use and the positive outcomes that will come with it (i.e. feeling less stressed).

Understanding the difference between outcome expectancies and efficacy expectations is very important when talking about social learning theory and substance abuse addiction. According to Bandura's (1977) social learning theory, "an outcome expectancy is an individual's estimate that a particular behavior will lead to certain positive outcomes" (Brandon et al., 2004, p. 54). An example of this would be a person believing that smoking cigarettes will help reduce their stress. Bandura explains an efficacy expectation as "the belief that one can successfully execute that behavior" (Brandon et al., 2004, p. 54). It is not necessary for the person's expectancies to be accurate in order to influence or trigger their behavior, but the person must believe that these expectancies are true. As long as a person believes that smoking cigarettes will reduce their stress, it is considered to be an outcome expectancy. It does not matter that cigarettes are not actually reducing a person's stress, because the individual believes it to be true.

After reading that the person's expectancies do not have to be true, but rather they just must be believed by the person to be true, I wonder how much a person can influence themself to believe that something is true even if deep down they know it is not. For example, if a person says that the reason that they drink alcohol is to help relieve the stress that they are experiencing in their life, but after they wake up from a night of drinking and all of the stress is still there to deal with, is this expectancy still something they believe to be true? If a person does believe certain positive expectancies to be true, such as drug use will help with weight loss or help alleviate stress, and then starts using and starts to see the negative consequences that are associated with their drug use or alcohol use, how often does that person start to change their mind about believing those expectancies?

In conclusion, social theory asserts that behavior is learned through direct behavioral conditioning, differential reinforcement, and through imitation or modeling of others' behavior. An individual who is around substance and alcohol abuse and sees it in a positive way with positive outcomes is more likely to engage in using those substances. On the other hand, an individual who is around substance abuse and alcohol abuse and sees the negative outcomes associated with it is less likely to engage in using those substances.

Based on my analysis of substance abuse through a social learning theory lens, I believe that one's environment and interpersonal relationships increase one's risk for substance abuse disorders. How we think affects the way we behave, and how we behave effects the way we think. According to social learning theory, the people we surround ourselves with are more important and influential than we may think!

References

Akers, R. L., La, G. A. J., Cochran, J., & Sellers, C. (December 01, 1989). SOCIAL LEARNING THEORY AND ALCOHOL BEHAVIOR AMONG THE ELDERLY. Sociological Quarterly, 30, 4, 625-638.

American Psychiatric Association. (2013). Diagnostic and statistical manual of mental disorders: DSM 5. Washington, D.C: American Psychiatric Association.

Brandon, T., Herzog, T., Irvin, J., & Gwaltney, C. (2004). Cognitive and social learning models of drug dependence: implications for the assessment of tobacco dependence in adolescents. *Addiction, 99*51-77.

Niaura, R. (January 01, 2000). Cognitive social learning and related perspectives on drug craving. Addiction (abingdon, England), 95, 155-63.

Peralta, R. L., & Steele, J. L. (January 01, 2010). Nonmedical Prescription Drug Use Among US College Students at a Midwest University: A Partial Test of Social Learning Theory. Substance Use & Misuse, **45, 6, 865-887**.

Racial Disparities in Diagnosing

Zakira Tenny

Introduction

Have you ever experienced being labeled and it influenced the way you felt about yourself? If the answer to this question is yes, you are reading the right paper. There are one out of four adults who are diagnosed with a mental health disorder as described in the *Diagnostic and Statistical Manual of Mental Disorders*, 5th Edition (National Alliance on Mental Health, 2013). In some cases, it is fine to be labeled if an individual has a mental illness that needs accessible resources and treatment. However, what does it indicate if an individual is being diagnosed with a mental illness that involves stigmatizing and institutionalized discrimination which puts a negative label on their well-being? Through this paper, the aim is to critically explore how diagnoses that are given to the African American population are questionable compared to their counterparts, Caucasian Americans. This paper will interpret the racial disparities that occur in the mental health system.

Theoretical Perspectives

Feminist Theory focuses on equality for both men and women in such a structuralized society. It is often assumed that feminist theorists are pro empowering women only, however feminists focus on a variety issues that are composed of hierarchies that create inequalities and discrimination (i.e. gender roles, gender stereotypes, sexuality, and more; Hawkesworth 2010). The purpose of using the Feminist Theory is to critically construct an argument that can offer insight to readers on the account of knowing that some diagnoses that are given are unjust and should be talked about more.

The Ecological Systems Theory will be used to speculate the different systems such as the Micro system, Meso system, and Macro System. Each system can play a role in how people are being diagnosed with one or more mental health disorders within the DSM 5. The micro system will focus on the person in their environment; the meso system will look at how the person is affected by his or her environment (i.e. community, school, etc.), and macro system will show how the law and/or rules have an impact on the person (Leonard, 2011). This theory is very vital when evaluating different diagnoses and their impacts on an individual's well-being.

Post-Traumatic Stress Disorder

As mentioned above, this text will be analyzing and critiquing how African Americans are being diagnosed from the DSM 5, and we will question whether the diagnoses given to this population are given fairly. In this section, the first and fourth criteria will be discussed. Post-Traumatic Stress Disorder (PTSD) is one of the many disorders that are located within the DSM 5, specifically under the Trauma and Stressor-Related Disorders section. PTSD is defined as the development of characteristic symptoms following exposure to one or more traumatic events (American Psychiatric Association, 2013, p. 274). The definition of PTSD is broad and could be misinterpreted by clinicians. However, the criteria discussed within this particular section give a better sense of what to look for in terms of diagnosing a patient. According to the DSM 5, emotional reactions are not used as a determinant of an individual having PTSD due to the reaction possibly having closely related symptoms as another disorder within the DSM 5 (American Psychiatric Association, 2013, p. 274). Each of the criteria discussed in this paper will be directly from the DSM 5.

Overall, the definition of PTSD can be perceived as a very broad and general mental health disorder. With each of the criteria listed under the PTSD section in the DSM 5, it gives brief insight of listed symptoms that are used to determine if the client should be diagnosed with the disorder. The first criterion also known as Criterion A describes the stressors of the disorder that a person should have experienced if they are suspected to have PTSD, such as: death, threatened death, actual or threatened serious injury, or actual or threatened sexual violence (American Psychiatric Association, 2013, p. 271). It is a requirement that an individual must have one of the following: direct exposure, witnessing, in person, indirectly (close relatives or friends exposed to trauma, if involved actual or threatened death, it must have been violent or accidental), and repeated or extreme direct exposure to aversive detail of events, usually in the course of professional duties. This criterion is only used to assess individuals aged 6 and older.

While this criterion seems to be very concise and well-put, the issue is primarily regarding the person who is performing the assessment on the pa-

---

tient. What is the true definition of actual or threatened sexual violence and how do clinicians/psychiatrists go about discovering it within each patient? For example, there are many instances where sexual violence (rape) goes unreported; does that indicate that the person is taken off the spectrum of post-traumatic stress disorder? According to a study by Boakye (2011), some rape ideologies seem to accept rape and deconstruct the notion of its incidence as negative and degrading. When there are people who accept rape, it often goes ignored due to implicitly blaming the victim. While the emotional reactions are not considered in the DSM 5, it is still important to recognize that the patients are feeling some form of emotion including helplessness, fear or horror, but it is possible that a person could experience one of the three instead of all (Resick & Miller 2009). For example, an individual could have experienced sexual molestation and may experience helplessness, but they do not experience any horror over the situation.

Another example that comes from a recent study regarding African American barriers, showed that many people who have experienced the symptoms of PTSD encounter financial barriers where there is no treatment available so they tend to avoid the situation (Davis et al, 2008). This is not unusual for low-income individuals which are a great concern due to having the knowledge that there are many people who are unable to seek the appropriate treatment. This can be an illustration of helplessness, which correlates with higher stress levels that an individual endures compared to middle-class counterparts.

The fourth criterion also known as Criterion D: negative alterations in cognition and mood (American Psychiatric Association, 2013, p. 271-272). A person must have two of the following symptoms: inability to recall key features of the traumatic event, persistent (and often distorted) negative beliefs and expectations about oneself or the world, persistent distorted blame of self or others for causing the traumatic event or for resulting consequences, persistent negative trauma-related emotions, markedly diminished interest in significant activities, feeling alienated from others, and constricted affect: persistent inability to experience positive emotions (American Psychiatric Association, 2013, p. 272). When looking at this specific criterion, it displays an abundance of symptoms that an individual must experience before being diagnosed. According to research, adolescents are often filled with guilt or feeling that they failed to do something in a situation where they had no control over (Kletter et al, 2009).

For example, there are many instances where African Americans in comparison to their Caucasian American counterparts have been exposed to close family and friends being murdered right before their eyes. "These youth are ten times more likely to be victims of homicide compared to their white peers, with homicide being the leading cause of death among African American youths aged 15-24" (Voisin, 2008). An increased amount of violence occurs within the African American populations compared to other segments; children are often exposed to it early on to the point of the child becoming numb to the situation. Although a person can try to dissociate the exposure of those violent memories, it is possible that it can be retrieved again in their memory. Knowing that violence has increased in the African American population, it would be appropriate for individuals to go through an assessment process with mental health services; however many are enduring financial barriers.

While the DSM 5 can be a useful guide to discovering the appropriate diagnosis for an individual, there are many minorities in the world who are not treated due to their socioeconomic status and what their medical card could afford. Many African American populations face several barriers regarding mental health access, and seek for help. According to the Hines-Martin et. al.((2003) study, barriers are recurring in this population ranging from the individuals' personal judgments toward mental health to the individuals' environment influencing access to mental health resources. Using the ecological theory, the micro system of an individual is their thoughts and perceptions of the situation, which could be more negative due to their having the notion that all mental health patients have a problem or are considered to be "crazy". This could be considered a barrier to an African American man/woman because of the negative connotation that is used widely implying that people who are diagnosed with mental health issues are not considered normal, so as a result they restrict themselves from getting the proper help.

Similarly to the micro system, the meso system greatly deters an individual from help seeking due to the distorted connoting opinions of their neighbors and community members. If a person is concerned about how others view them, it is likely that they will not seek the help needed to refrain from being stigmatized. A person's environment can have an impact on their decision making which leads me to the macro System. The macro system in this case is very crucial with analyzing the differentiated medical treatment in the minority populations compared to other segmenting populations. A qualitative research study involving African Americans and their access to mental health services found that there are a variety of institutionalized barriers that are hindering these individuals from receiving the appropriate health care such as, specific time and limitations, gatekeepers, and rules (Hines-Martin et al, 2003).

Health care can be very expensive around the world, and patients who have Medicare or Medicaid are often faced with the insurance paying a little portion and leaving the remaining balance with the patient who can barely afford to pay the balance (Burgess, 2007). This indicates that many minorities face greater financial challenges compared to their middle-class counterparts. The system is seemingly structured with intent to have the minorities struggle due to setting out specific limitations in places where people need the most assistance. All of these factors could have an influence with one's thoughts of guilt and shame, so the remaining question that lingers is the PTSD being diagnosed sufficiently across the racial spectrum of African Americans and Caucasian Americans?

Autism Disorder

Autism spectrum disorders (ASD) are characterized by severe and pervasive impairments in three domains: reciprocal social interaction and communication skills and the presence of stereotyped behavior, interests, and activities (Chloemkery et al, 2013). ASD can be associated with Conduct Disorders. The first three criteria will be examined closely in this particular section in order to differentiate the two. The first criterion insists that patients with autism should experience persistent difficulties in the social use of verbal and nonverbal communication as manifested by one of the four symptoms that are listed in this section (American Psychiatric Association, 2013, p. 50). For example, patients who are autistic suffer from abnormal social behaviors such as social and emotional reciprocity, social interaction, and maintaining and understanding relationships (American Psychiatric Association, 2013, p. 50). In this particular category of symptoms it goes in greater detail with specific examples that are helpful to understand what the professionals should look for. However, it can be very difficult to diagnose people with such disorders due to it to being closely associated with other behavioral disorders. ASD can be associated with a form of behavioral disorders such as Conduct Disorder (CD) considering the symptoms that are listed above.

African Americans compared to Caucasians are often diagnosed with more behavioral disorders rather than disorders linked to Autism Spectrum disorder (Clark, 2007). There are some instances where a person can repeatedly display behaviors that are against the social norms through their interactions, but where is the line drawn into determining if a person should be diagnosed with a behavioral disorder or autism spectrum disorder? This becomes very questionable about how often this occurs and its rationale. Could race be an issue in this case? For example, people who are diagnosed with autism are usually given interventions and some form of counseling, while on the other hand those that are diagnosed with behavioral disorders are being associated to delinquency (Clark, 2007). Using the Feminist theory, this displays a form of inequality that is often used through institutions with the people barely recognizing the problem of certain diagnoses. Through a feminist lens, the goal to overcome this challenge will be through social policy and discovering ways that can increase the likelihood of unraveling barriers that minority communities endure (Coady& Lehmann, 2008, p. 344-345).

The second criterion of ASD is restricted, repetitive patterns of behavior, interests, or activities, as manifested by at least two of the following, currently or by history: Stereotyped or repetitive motor movements, insistence on sameness, inflexible adherence to routines just to name a few (**American Psychiatric Association, 2013, p. 50**). The issue that has risen in this criterion is the broad and generalizable wording which can cause a sense of misconception. For example, the criterion does not explicitly lay out the expression of stereotyped or repetitive motor movements. What does that look like? It is very important to diagnosis individuals on an individualized basis rather than generalizing the symptoms with previous patients that have already been diagnosed. If people are being diagnosed from a generalized perspective, it can possibly lead to patients being misdiagnosed.

From an ecological perspective, some symptoms that are highlighted in the DSM V can be caused genetically or environmentally (**American Psychiatric Association, 2013, p. 51**). African Americans who may possibly show signs of the symptoms could be influenced by their direct and indirect environment. For example, African Americans are sometimes exposed to different environments which can have effects on their behaviors. Minorities living in low-income areas are often secluded from resources that are made more available to white middle-class counterparts (Tek&Landa, 2012). This leads them to not getting the appropriate help needed. Behaviors can also be learned in the home, which could resemble the symptoms of this criterion, however if the same kinds of symptoms are being shown in African Americans as well as Caucasians, how are the diagnoses being diagnosed differently?

There is a disparity that is ongoing across ethnicities with the different diagnoses that are being made between ASD, CD, and other behavioral disorders (i.e. adjustment disorder, ADHD). Recent research shows that African Americans are much more likely to receive CD or adjustment disorder compared to Caucasian counterparts (Mandell et al, 2007). This indicates that there is implicitly a form of discrimination that tends to occur across the mental health spectrum regarding African Americans and other minorities. The diagnoses of Autism Spectrum Disorder should be closely evaluated pertaining to the determinants of diagnosing minorities versus non minorities. In the end, being misdiagnosed can consequently have an influence on a person's life/and or future.

Conclusion

It has been shown that there are a variety of misdiagnoses that occur in the mental health care systems regarding African Americans. Although there were only two disorders from the DSM V mentioned in this paper, it is important to gain the awareness that any of the disorders out there could experience similar disparities. Institutionalized discrimination is highly relevant in this case due to the disorders that are being diagnosed in African Americans are negatively connoted in the world stigmatizing these groups through the diverse diagnoses (Burgess et al, 2008).

Having these negative connotations of bad behavior and disorderly social interactions can create barriers in African Americans excelling in society. When you look at how expensive health care is in the United States, it is ridiculously hard for someone who needs assistance with their disability to get the resources that are needed due to financial instability. However, White middle-class Americans are able to have the advantage of medical health care due to it being more accessible in their communities.

The DSM V is made to guide clinicians in diagnosing patients and providing them with the proper resources that are needed. When the issue of racial disparities within mental health care arises, the DSM V, along with some clinicians, develops contradictions concerning the intentions of such diagnoses.

There needs to be more research on this topic due to it lacking the respectable publicity to inform the people on issues such as racial disparities. This will assist with opening the opportunity for change and equality.

References

American Psychiatric Association. (2013). *Diagnostic and statistical manual of mental disorders* (5th ed.). Washington, DC: American Psychiatric Association.

Boakye, K. E. (2009). Attitudes toward rape and victims of rape: a test of the feminist theory in Ghana. *Journal of Interpersonal Violence*, 1633-51.

Burgess, D. J., Ding, Y., Hargreaves, M., van, R. M., & Phelan, S. (2008). The association between perceived discrimination and underutilization of needed medical and mental health care in a multi-ethnic community sample. *Journal of Health Care for the Poor and Underserved* , 894-911.

Cholemkery, H., Kitzerow, J., Rohrmann, S., & Freitag, C. (2014). Validity of the social responsiveness scale to differentiate between autism spectrum disorders and disruptive behaviour disorders. *European Child & Adolescent Psychiatry*, 81-93.

Clark, E. J. (2007). Conduct Disorders in African American Adolescent Males: The Perceptions that Lead to Overdiagnosis and Placement in Special Programs. *Alabama Counseling Association Journal*, 1-7.

Coady, N., & Lehmann, P. (Eds.). (2008). *Theoretical perspectives for direct social work practice*. 2nd ed. New York, NY: Springer.

Davis, R. G., Ressler, K. J., Schwartz, A. C., Stephens, K. J., & Bradley, R. G. (2008). Treatment barriers for low-income, urban African Americans with undiagnosed posttraumatic stress disorder. *Journal of Traumatic Stress*, 218-222.

Hawkesworth, M. (January 01, 2010). From Constitutive Outside to the Politics of Extinction: Critical Race Theory, Feminist Theory, and Political Theory. *Political Research Quarterly*, 686-696.

Kletter, H., Weems, C., & Carrion, V. (2009). Guilt and Posttraumatic Stress Symptoms in Child Victims of Interpersonal Violence. *Clinical Child Psychology and Psychiatry*, 71-83.

Leonard, J. (2011). Using Bronfenbrenner's Ecological Theory to Understand Community Partnerships: A Historical Case Study of One Urban High School. *Urban Education*, 987-1010.

Mandell, D. S., Ittenbach, R. F., Levy, S. E., & Pinto-Martin, J. A. (2007). Disparities in Diagnoses Received Prior to a Diagnosis of Autism Spectrum Disorder.*Journal of Autism and Developmental Disorders*, 1795-1802.

National Alliance on Mental Health Illness. (2013). *Mental Illness Facts and Numbers. Retrieved from:* http://www2.nami.org/factsheets/mentalillness_factsheet.pdf

Resick, P. A., & Miller, M. W. (2009). Posttraumatic stress disorder: Anxiety or traumatic stress disorder?. *Journal Of Traumatic Stress*, 384-390.

Tek, S., & Landa, R. J. (2012). Differences in Autism Symptoms Between Minority and Non-Minority Toddlers. *Journal of Autism and Developmental Disorders*, 1967-1973

Voisin, D. R. (2008). The Effects of Family and Community Violence Exposure among Youth: Recommendations for Practice and Policy. *Journal of Social Work Education,* 51-66.

Should Kleptomania Really
Be Considered a Mental Disorder?

Emily C. Vanella

Introduction

The Diagnostic and Statistical Manual of Mental Disorders (Fifth edition), or DSMV as it will be referred to for the duration of the paper, is a good resource for professionals who deal with clients who suffer from mental disorders. It is used to pinpoint what disorder the client and clinician are working with. All of the diagnoses seem fairly straightforward, except for one, kleptomania, which by definition could be confused with shoplifting. This chapter will discuss the definition of kleptomania and the definition of shoplifting to try to decide if they are one and the same, and therefore should not be considered an actual mental disorder. This chapter will also discuss cognitive- behavioral theory and discuss how the many models that are associated with cognitive-behavioral theory perceive kleptomania.

There have not been a lot of studies conducted on kleptomania, so the author will use cognitive-behavioral theory as a lens to show that the thoughts and behaviors of someone with kleptomania are the same as someone who is labeled as a shoplifter. Some mental health practitioners and experts say cognitive- behavioral theory can be used to help treat kleptomania and the co-morbid disorders that are often associated with kleptomania. Finally, the author will conclude with an argument that kleptomania should be considered as a possible accompanying disorder of other disorders, and not in itself a primary disorder.

Kleptomania

According to the DSMV, the criteria for an individual to be diagnosed with kleptomania include failing to resist impulses to steal objects that are not needed for personal use or for their monetary value on multiple occasions. It is important to note that the DSMV allows "multiple occasions" to be interpreted by the clinician; does this mean it has to happen at least two times? Or does it have to happen more than two times? Do the occasions occur on different dates or could they be multiple thefts during the same day? Do the incidences have to happen at different locations/ stores? Individuals experience an increasing sense of tension immediately before committing the theft, and pleasure, gratification, or relief once the theft has been committed. The act of stealing is not committed to express anger or vengeance, and is not in a response to a dare, as an act of rebellion or right of passage. The act of stealing can also not be in response to a delusion or hallucination. The act of stealing is not better explained by conduct disorder, a manic episode, or antisocial personality disorder. Individuals with kleptomania typically attempt to resist the impulse to steal and are aware the act is wrong and senseless. They fear being apprehended and often feel depressed or guilty about the thefts (American Psychiatric Association, 2013, p. 478).

From the biological and neurological perspective it is important to take note that neurotransmitter pathways associated with behavioral addictions, including those associated with serotonin, dopamine, and opioid systems, appear to play a role in kleptomania as well (American Psychiatric Association, 2013, p. 478). This is important to take note of because if the client or clinician decides to take pharmaceutical action to help treat this disorder, it is important to understand what hormones are affected. It is also important to keep in mind when looking for more information or studies to help understand this disorder to look at studies which focus on neurotransmitter pathways.

The DSMV estimates that 4-24% of people arrested for shoplifting have kleptomania. This implies that there is a difference between ordinary theft and stealing because of kleptomania. Ordinary theft is deliberate and is motivated by usefulness of the object or its monetary worth. Other differential diagnosis includes malingering, which is when individuals simulate the symptoms of kleptomania to avoid criminal prosecution for shoplifting. Individuals with antisocial personality disorder and conduct disorder show a general pattern of antisocial behavior, which can include stealing (American Psychiatric Association, 2013, p. 478).

Kleptomania is seemingly very rare, and the DSMV predicts that 0.3-0.6% of the general population has kleptomania. Women are more likely than men to have kleptomania, and it is generally found in adolescents and young adults, but can also be found in older adults. Kleptomania may also be associated with compulsive buying as well as with depressive and bipolar disorders. Other DSMV disorders which may accompany kleptomania are anxiety disorders, eating disorders particularly bulimia nervosa, personality disorders, substance use disorders especially alcohol use disorders and other disruptive, impulse-control and conduct disorders (American Psychiatric Association, 2013, p. 478-479).

Shoplifting

Shoplifting in its simplest terms is the act of knowingly obtaining goods from an establishment in which they are displayed for sale without paying for them. Shoplifting is often not a premeditated crime. In fact, 73% of adult and 72% of juvenile shoplifters do not plan to steal in advance. There are approximately 27 million shoplifters (or 1 in 11 people) in the United States today. More than 10 million people have been caught shoplifting in the last five years. Most of these shoplifters have done it more than once, though previous times they may not have been caught (Leary, 2013; Egan and Taylor, 2010).

Martin S Humphrey wrote an article for *The British Journal of Psychiatry* that summed up shoplifting in 100 words:

> Shoplifting is common. The Centre for Retail Research estimates that customer and staff theft in the UK accounted for nearly £4000 million (which is just under 6000 million US dollars) in 2011 alone. Generally there is no link with mental disorder. But it can be associated with depression, often in apparently well- adjusted, law- abiding women, in middle or later life and what appears to be a conventional marriage. Characteristically in such cases it occurs in a major high street retailer, is of an unwanted item that the perpetrator possesses the funds to buy, which is removed if not ostentatiously, then with no attempt at concealment (Humphreys, 2013).

Humphrey's article is really just a formal definition of kleptomania. Some may argue that the difference between the two is that kleptomania is a type of impulse control disorder. So what is the driving force for over 70% of shoplifters if it is not premeditated, if not an issue with impulse control?

Cognitive-Behavioral Theory

Cognitive-behavioral theory, or CBT as it is commonly referred to, is constantly growing and changing as time goes on. Its primary author, Aaron Beck, combined cognitive therapy and behavioral therapy. During the 20th century cognitive therapy had a lot of success, but it had little effect on those affected with depression. Beck noted that a lot of the distress that came from depression was not due to the unconscious, as Freud had predicted, but rather from automatic thoughts that seemed to be reinforced by negative behaviors (Trull, 2007; Purdon, 2004).

Cognitive-behavioral theory has a few main ideas: cognitive activity affects behavior, cognitive activity may be monitored and altered, and behavior change may be achieved through cognitive change. Cognitive-behavioral theory combines cognitive, behavioral, psychoanalytical, sometimes biological, and strengths perspective models (Kohn, 2006). Cognitive- behavioral theory and therapy uses many models to help clients change the way they think about situations, which will in turn "correct" behaviors and emotions.

Cognitive-behavioral theory is most popular for treating depression and anxiety. Depression and anxiety are frequently found in all types of obsessive-compulsive and impulse control disorders (Purdon, 2004). .

Through the Lenses of CBT

Using cognitive- behavioral theory to analyze kleptomania focuses on the first idea of the theory to analyze those individuals' actions. Cognitive activity affects behavior. It would look like this: An individual is in a retail store. The individual experiences or thinks of a trigger of some sort, which makes that individual anxious and distressed and with a feeling of dread like something bad is going to happen. The individual automatically, almost unconsciously, thinks "if I take this object/good home with me I will feel better, this distress or feeling of dread will go away." The thoughts happen so fast that the individual is almost unaware that the thought is occurring. Then they sometimes feel anxious on top of the unknown anxiety that they may get caught. And when the act is completed, the feeling of dread and anxiety is gone, and they feel relief. The same cognitive pattern is found for impulse shoplifters who have no other motive other than they just did it. Again that is about 70% of individuals caught shoplifting do so with out premeditative motives.

Individuals who are labeled as shoplifters and do not do it on impulse but are motivated by other reasons, such as other psycho- social factors like necessity, or social pressures (such as a dare), still have feelings of distress due to shoplifting. Those individuals can think of psycho- social factors constantly, and may become depressed or anxious about them. They think that the only way to stop those unpleasant experiences and emotions is to steal; the only way they are going to get those necessities or get the social approval is to steal. The only difference for a cognitive- behavioral theorist is those thoughts of worry, anxiety, and depression are in the front in the individual's mind. They are conscious of the thoughts and use those thoughts as the reason and motivation to steal. The cognitive activity still affects the behavior of the individual to steal.

Treatment Using CBT

Professionals who use cognitive-behavioral theory see understanding the theory is a way of treatment if used properly. Cognitive- behavioral therapy is the theory put into action to stop maladaptive behaviors by changing and challenging the thoughts of the individual which lead to said behaviors. In other words it uses the second and third ideas of the theory as the guidelines for the therapy; cognitive activity may be monitored and altered, and behavior change may be achieved through cognitive change. So the correct way to stop someone from stealing is to change the way they think in order to change the behavior of stealing. As mentioned before cognitive-behavioral theory/ therapy can be used in conjunction with other common methods such as cognitive, behavioral, psychoanalytical, biological, and strengths perspective models if the professional helper thinks it would be beneficial for the individual who wants to change the maladaptive behavior.

It should be noted that because a behavior can, should, or needs to be changed; and it can be done by a psychological method such as cognitive- behavioral theory, does not mean that individual possessing said behavior has a mental disorder.

Kleptomania Controversy

According to Carolynn Kohn, kleptomania is generally an unknown and un-researched topic (2006). Professional views widely differ from treating Kleptomania as an affective spectrum disorder to more of an obsessive-compulsive disorder. If looked at as an obsessive- compulsive disorder, the criteria are very similar, especially when looking at the thought processes of someone diagnosed with obsessive- compulsive disorder and someone diagnosed with kleptomania. Kohn stated, "What is clear, however, is that symptoms of kleptomania rarely occur in isolation, and frequently occur in conjunction with other mental health problems" (Kohn, 2006). This being said, how can professionals be sure it is a separate diagnosis and not a symptom that can occur with the other mental health problems? Are said other mental health problems actually the primary diagnosis? There have been no studies to show that kleptomania is truly and completely a primary diagnosis. It is almost always associated with another diagnosis, and when it is on its own it can easily be dismissed as shoplifting.

According to the DSMV, the main difference between shoplifting and kleptomania is that kleptomania disrupts normal functioning in an individual's life. This means that for an individual to have a mental diagnosis of kleptomania, the compulsion to steal causes legal, family, career, and personal difficulties and chronically disrupts the functioning of the individual's life (American Psychiatric Association, 2013, p. 479).

The issue with requiring a disruption to daily functioning is that anyone caught and prosecuted for theft will by definition have legal difficulties. Being convicted of a crime will almost always result in career difficulties, and would most likely disrupt the individuals' family and personal life; all making it difficult to sustain a normal life.

Another problem with the way the DSM discusses kleptomania is the predicted percentages of those diagnosed with it. The DSMV considers kleptomania to be a very rare diagnosis and estimates that 4-24% of arrested shoplifters have kleptomania, while only 0.3 - 0.6% of the general population has kleptomania (American Psychiatric Association, 2013, p. 479). How can the percentage of arrested shoplifters who have kleptomania be so low when other studies show about 72% - 73% of shoplifters (Leary, 2013) seem to meet the requirements of kleptomania? Either the definition of kleptomania is not specific enough or those millions of people all have un-diagnosed kleptomania.

How to Fix the Kleptomania Diagnosis

There are a few actions the American Psychiatric Association, authors of the DSMV, can take to improve the diagnostic criteria of kleptomania. One solution is to take kleptomania out of the DSMV altogether. If kleptomania needs to be in the DSMV, it should be included as a subtype of obsessive-compulsive disorder or as possible criteria for other diagnoses.

Another possibility is to do more research on the subject. This should be done regardless of which corrective action, if any, is taken. Today the studies on the subject of kleptomania are few and far between, and the research gathered needs to be used to make the criteria more specific. "Multiple occasions" specifically in the criteria needs to define "multiple" more specifically. The DSMV is fairly specific with other disorders such as delusion disorder where the first criteria is suffer one or more delusions with a duration of one month or longer (American Psychiatric Association, 2013, p. 90). The DSMV's obtuseness for the kleptomania criteria makes it difficult to diagnose.

Say an individual, Mr. A, could not control the impulse to take something that he did not pay for. Mr. A has only done this on three separate occasions over the last 35 years; Mr. A did this with no coercion or as a response to a hallucination or delusion and had no real need for the items he took. Now lets say Mr. B could not control the impulse to take something every single time he walked into a store. Mr. B has only felt the need to do this for the past 2 weeks. Mr. B has no need for the items he has taken, and does not take them in response to a delusion or hallucination or any sort of coercion. Would Mr. A and Mr. B both be diagnosed with the same mental illness of kleptomania?

If the American Psychiatric Association feels the need to keep kleptomania as a mental disorder, and assuming they find more evidence/ research on it, they will need to update the percentages of those with the disorder. Or at least tell readers where they are getting their data/ percentages. The lack of specification seems to be another result of a confusing criterion.

Conclusion

After looking at the DSMV's criteria, definition, and statistics of kleptomania, compared to the criteria, definition and statistics of shoplifting, and examined through the lends of cognitive-behavioral theory, the obvious conclusion is that kleptomania should not be in the DSMV as is. If kleptomania is a true mental disorder there needs to be more research on it and more specified criteria; otherwise there would need to be diagnoses of every shoplifter who acted more than once, and had no motive as a kleptomaniac. Otherwise kleptomania should be a possible sub-criterion for other disorders. Whether kleptomania is an actual mental disorder or just a bad behavior, it is better understood through the means of cognitive- behavioral theory and can be altered by cognitive- behavioral therapy.

References

American Psychiatric Association. (2013). Diagnostic and statistical manual of mental disorders: DSM 5. Washington, D.C: American Psychiatric Association.

Humphreys, M. S. (2013). Shoplifting -- in 100 words. *British Journal of Psychiatry*, 202(2), 128-128.doi:10.1192/bjp.bp.111.100750

Kohn, C. S. (2006). Conceptualization and treatment of Kleptomania behaviors using cognitive and behavioral strategies. *International Journal of Behavioral Consultation & Therapy*, 2(4), 553-559.

Leary, Mark.(n.d.). *Why people take part in violent flash mobs*. Duke University News and Communications.

Purdon, C. (2004). Cognitive-behavioral treatment of repugnant obsessions. *Journal Of Clinical Psychology*, 60(11), 1169-1180

Trull, T. J. (2007). Clinical psychology (7th Ed). Belmont, CA: Thomson/Wadsworth.

Egan, V. & Taylor, D. (2010) Shoplifting, unethical consumer behavior, and personality. *Personality and Individual Differences, 48(8)*, 878-883. doi: 10.1016/j.paid.2010.02.014

Understanding Posttraumatic Stress Disorder through Systems Theory

Stephanie Wessels

The *Diagnostic and Statistical Manual of Mental Disorders* (DSM) is a catalog of symptoms associated as criteria to different diagnoses. However the application of a specific perspective can shift how that diagnosis is understood. Systems theory provides a unique insight into the complexities of posttraumatic stress disorder as a diagnosis and experience. It exposes the perpetuation of trauma, as well as the vastly different understanding of what can consist of a traumatic event. Systems theory provides a lens to better understand: the interrelatedness of individuals and systems; how culture and identity, as well as relational experiences and context, play a role in a person's experience of stress; and the extent of secondary trauma exposure. This exploration is not to belittle those who meet the criteria for posttraumatic stress disorder as it stands. Any experience of trauma in people and its disruptive and potentially harmful nature is valid. However, there is a broader disruption that occurs. Implications of this may contribute to the United States of America's disproportionate rate of posttraumatic stress disorder compared globally.

Posttraumatic Stress Disorder as a Diagnosis

Posttraumatic stress disorder (PTSD) was officially included by the American Psychiatric Association (APA) in the *Diagnostic and Statistical Manual of Mental Disorders'* third edition in 1980 (Basham, 2011). The reported and described trauma responses did not miraculously become prevalent in the 20th century, but instead have been documented throughout time, particularly around war related combat experience. PTSD's induction as a formal diagnosis was the result of advocacy by Vietnam Veterans who were seeking support and recognition of their struggles from the surrounding community (Basham, 2011, p. 442). In the DSM 5, the APA maintains PTSD as a diagnosis, within the Trauma- and Stressor-Related Disorders category.

The diagnostic criteria are separated into eight primary criteria that the DSM 5 outlines alphanumerically. Criterion A calls for direct exposure to a traumatic event. This factor is particularly unique, as this diagnosis calls for a clear external event as a precursor to the symptoms. Symptoms are clustered in criteria B and C as intrusion/re-experiencing symptoms and avoidance symptoms respectively, calling for at least one symptom present (APA, 2013). Criteria D and E are grouped in negative cognitions and mood symptoms and symptoms of hyper-arousal respectively, requiring two symptoms present in each criterion (APA, 2013). Other requirements include the "disturbance causing clinically significant distress or impairment in social, occupational, or other important areas of functioning" for longer than one month, while not being attributed to effects of substance use, other medication or another medical condition (APA, 2013, p. 272). If the individual is younger than six years old, some specific symptoms are not applicable, but the general clusters remain consistent.

As established, the diagnosis of posttraumatic stress disorder is relatively new to formal academic thinking. It also uniquely requires an external event and has roots as a formal diagnosis from sociopolitical influences. Vietnam veterans, feeling stigmatized and marginalized, organized to advocate for recognition of PTSD as a diagnosis (Bride, 2012). Those roots have an inherently limited scope, which can carry implications for the progression of the diagnosis. Although the baseline of trauma and context for trauma may have been founded with that particular circumstance in mind, it can and did expand to include a broader scope. As such, it is understandable that there may be significant limitations to the initial application.

Although PTSD is a relatively new diagnosis, the human experience of trauma is certainly nothing new. It is described in clear and specific terms in the DSM 5, but trauma is actually much more complicated. External events are nuanced and deeply entrenched in a person's individual experience and including his/her relation to other people and outside forces. These outside factors can be understood as systems. The sociopolitical context for the formalization of PTSD and its subsequent limitations is an example of how systems impact an individual object.

The American Psychiatric Association (2013) reports that while the "prevalence of PTSD is about 3.5% in the United States, lower estimates are seen internationally near 0.5%-1.0%". This drastic difference does not imply that Americans are more traumatized, but perhaps have different relationships to formal diagnosis and how trauma is understood culturally. The relationships between people and their environment map systems that inform how the presence of PTSD is or is not understood. To best understand this interplay and impact, systems theory should be more clearly discussed.

Theoretical Perspective: Systems Theory

Systems theory is intrinsically present in social work in many ways. The field's focus on individuals within an environment sets the stage for systems theory to be adapted as a general perspective. Early on, social work's focus was on individuals and families equally, acknowledged the importance of people's social contexts outside of the [counseling] office (Rothery, 2008, p. 90). Although systems theory can appear intuitive, there are guiding principles that help formalize it.

A system is made up of elements; those can be people, groups of people, ideas, environments, etc., and all of these different elements influence one another. These relationships tend to have a circular flow of causation and expose the intricate interrelatedness of people (Rothery, 2008, p. 91). Subsystems exist as groupings of elements within larger systems, and can be different depending on the focal element. So for instance, within a family system the parents are subsystem in relation to an individual child. Additionally, each element plays a particular role within the system and has primary functions as a result. Different roles and elements provide various resources to the larger system. To illustrate, a subsystem of parents may have different duties within the family system, like financial contribution and child-rearing. While understanding systems can provide valuable information, it does not map a perfectly clear picture of how each element relates. So although a child may come to expect financial support from her parents, that does not communicate all the reasons why or how she has come to that conclusion. Ultimately, systems often behave somewhat predictably as a result of the established structure. Even though many pieces can be nuanced, the picture formed is generally stable and consistent in message. A family system will continue to behave a similar way under similar conditions so that the elements within can predict and function inside it.

Patterns of behavior and boundaries form to make systems manageable and useful to those a part of it. Boundaries can be understood as invisible barriers between systems and/or elements (Robbins, Chatterjee & Canda, 2012). How the boundaries form and function depend on the specific elements. For example, a child may share different kinds of information with a subsystem of his/her parents as opposed to a subsystem of his or her siblings. That child is also impacted by a variety of environmental factors that can influence the types of boundaries and how she understands those boundaries as well.

A system adjusts and responds to maintain balance, a process called homeostasis, and does so regardless of a positive or negative outcome (Robbins, et al., 2012). This can lead to certain systems becoming more specialized and differentiated over time (Robbins, et al., 2012). Correlation here is still limited, and narrative histories should be considered vital components of an individual's context. Ultimately, human systems are best understood holistically where the whole is more valuable than the sum of individual elements. Systems theory provides a framework to evaluate specific aspects, and as Robbins, et al., (2012) mentions, it can eliminate assumptions of pathologies and remove the medical model approach. The medical model assumes that a presentation of symptoms necessitates treatment. Looking at posttraumatic stress disorder through systems theory can counter the reductionist assumptions and broaden the scope of what a diagnosis of PTSD represents.

PTSD and Systems Theory: A Critique

Systems function in an interrelated and circular way that can be difficult to describe in western language. This contradicts the linear progression somewhat unique to the diagnosis of PTSD enacted by the requirement of an external event. PTSD symptoms are commonly viewed as having a negative impact on a system's ability to function along with the negative impacts on the individual's ability to cope (Drozdek, 2012). However, when using systems theory to approach this particular diagnosis we recognize that trauma anywhere in the system may cause a multidirectional effect and will be felt and absorbed by the entire system.

This challenges the reductionist nature of posttraumatic stress disorder as a mental health diagnosis, where a presence of symptoms reduces an individual pathologically to one diagnosis down an invisibly direct path. Although some may summarize it as a "disorder of failed recovery" (Kirkpatrick & Heller, 2013, p. 343), it can also be more positively understood as a reorganization of coping skills as a result of a perceived severe stressor. The systems theory principle of equifinality is the concept that many different starting points and paths may ultimately lead to the same outcome, which directly contradicts the reductiveness of the diagnosis. Although a person may experience a traumatic event followed by significant distress from symptoms outlined in the DSM 5, it does not mean those things are causally related. The simplicity of the diagnosis of posttraumatic stress disorder does not adequately recognize the role of equifinality, specifically within: the role of interrelation; cultures and identities as systems; relational experiences and context; and exposure to secondary trauma.

Interrelation and Fit

While considering systems theory, there are various specifics within the general theory, one being the ecological approach. Foundationally it evaluates whether or not the person and environment fit well together. Systems theory understands the individual as deeply embedded within and amongst her environment. Germain and Gitterman write that systems theory "...defines the person and environment fit as the actual fit between an individual's perceived needs, goals and capacities and the responsiveness and quality of the person's physical and social environment within a historical and cultural context" (as cited in Acevedo & Gonzalez, 2011, p.165). A diagnosis of posttraumatic stress disorder does not always necessarily imply that a person is deficient, but rather that his or her environment may be deficient and potentially harmful. However, just because the needs, goals and capacities may have changed, it does not imply the system cannot respond appropriately.

Systems change, adapt and respond to various inputs and outputs. Harvey (2007) describes the power and equifinality of social contexts in three broad factors:

> a) Relationship dimensions include such attributes as participants' support of one another and the degree of spontaneity and open expression among them, b) personal growth and goal orientation dimensions include the extent to which the context provides opportunities for personal growth, c) systems maintain and change dimensions include qualities such as clarity of purpose and responsiveness to change. (p.16)

Systems have the power to be both harmful and helpful in each experience depending on the interpretive understanding of the individual. This understanding can be reasonably expanded to include experiences or trauma that meet the criteria of posttraumatic stress disorder. Although there may be a clear traumatic event, other factors can impact symptom expression. Furthermore, the reverse can also take place, meaning – even if there is not a clear traumatic event the symptoms root may be a different form of trauma.

Culture and Identities as Systems

A person's culture as well as their own personal identity function as systems. Those types of systems may often operate discretely, including in informing and interpreting posttraumatic stress disorder symptoms. However, PTSD is widely criticized for a lack of cross-cultural application. Mather (2012) proposes that a unique narrative can frame the impact of trauma within an environment. Culture is an ever-changing system that will continue to shape interpretation and adaptation to the consequences of trauma. Micale asserts that Western society ascribes a pathological view of posttraumatic responses (as cited in Evans and Coccoma, 2014); therefore, non-westerners or people who do not view the pathological piece to those responses may be secondarily victimized by systems that impose an opposing view.

Other cultures often have their own "…healing approaches that draw upon indigenous storytelling, dance, spiritual rituals, and community building to alleviate trauma-related suffering" (Basham, 2011, p. 454). Even the prescription to alleviate the suffering is a Western cultural system interpretation of the response. Culture vastly impacts the understanding of trauma and appropriate responses depending on the individual. As a result the diagnosis may become harmful to the individual.

Like cultural identity, race and gender can also be impactful. Racial identities form over time and are carried through history in innumerable ways. Within the consideration of trauma, Miehls (2011) highlights the experiences of collective trauma and perceived discrimination – calling for an expansion of the PTSD criteria to include this complex, continual view of trauma. On the surface, gender appears to the primary factor in the prevalence of the diagnosis. Statistically more women experience trauma that results in an official diagnosis; systems theory implies that a simplistic reduction is inadequate. Mather (2012) reports that high rates of sexual victimization among women contributes to PTSD being more prevalent in this population. Consequences impacting the prevalence of that sexual victimization are vast and numerous among a variety of systems. Personal identity can be represented by things beyond culture, race and gender. An individual can identify with multiple identities. Again, PTSD's narrow focus contradicts the interrelatedness of humans.

Individual Relational Experiences and Context

Systems relate to each other and to individuals in different ways. Those relational experiences represent the individual's agency and exercise of control. The diagnosis of PTSD's requirement for an external event usurps some of that control from the individual and how he/she may interpret it. Van de Kolk (2002) highlights the damage that may happen from differing opinions of trauma by explaining that a "...lack of validation and public acknowledgement, such as usually occurs after an attack by acquaintances, tends to lead to shame, helplessness, secrecy and preoccupation with maintaining one's emotional connections and financial security" (p. 382). These reactions may happen within and amongst a system as well. For example, parents may not acknowledge or recognize the gravity of a trauma their child experiences. Or a family can experience trauma in part as a result of systemic racism. Those outcomes may be interpreted as symptoms as well, but perhaps they are just mistreatment of systems.

Ultimately, occurrences of trauma shape how systems organize themselves, and because of the nonlinear nature of PTSD, the systems response is just as valuable in the evaluation of the trauma itself. Individuals and systems have a reciprocal nature as well and hold the power to create and/or compound trauma for others. An individual's ability to remain a part of the system or being forcibly or inadvertently removed also impacts the way PTSD is understood. Systems implication continues to extend beyond the relational experience and also should account for the context in which the event takes place.

The breadth of traumatic events is immense and what may be disruptive to one person may not neatly fit within the criteria outlined in the DSM 5. The current symptom criteria do not account for what is considered complex or collective trauma. Basham (2011) discusses the "intergenerational transmission of legacies of trauma" (p. 442). As it stands the DSM 5 recognizes that PTSD symptoms can be disruptive, but does not consider the damage that other's interpretations and actions impose on the impact of the traumatic event. Additionally the role of the individual's personal experience and role within systems impact the view of trauma. Different trauma and interpretations can continue to be transmitted through interfacing systems and people with which the individual is in contact. A child may experience trauma within the context of a family system, but her outside supports such as school, extended family, friends, etc., will form the reality within which the trauma is evaluated.

Secondary Trauma Exposure

Singling out posttraumatic stress as a diagnosis for one person also has implications for systems. It may open related and connected systems to secondary trauma and further transmission of trauma. Basham (2011) suggests that "...secondary traumatic stress and secondary trauma involves the natural behavior and emotions resulting from knowing about the traumatic events experienced by another person" (p. 445). The possibility of secondary trauma is deeply entrenched in the most basic role systems play. Looking amongst family systems can help illustrate the potential intrusion of trauma.

People within family systems can share symptoms though the trauma or a precipitating event in a variety of ways. Kilic, et al., reports "symptoms displayed by a symptomatic family member can increase the risk for secondary traumatization of children or other family members" (as cited in Bernardon and Pernice-Duca, 2010, p. 351). Kilic went on to clarify that specific symptoms of PTSD, like avoidance, may contribute to bad parenting styles that may negatively affect the children. Specifically symptoms like avoidance or hyper arousal behavior may cause the parent to overlook children's anxieties or symptoms of their own. Therefore secondary exposure can inflict primary trauma. This realization can also compound collective trauma in a continually covert way. Although some research has been done in regards to the role of trauma within family systems, it is acceptable to extend some of these generalizations to other kinds of systems. How these systems interact and communicate continue to play a significant role in how a diagnosis can be recognized and assessed.

Compassion fatigue. The impact of trauma is not discriminatory amongst types of people or the roles they play. There can be detrimental effects impacting individuals in caretaker roles of those primarily experiencing the trauma. Figley defines compassion fatigue as "exhaustion from long-term emotionally demanding situations" (as cited in Evans & Coccoma, 2014, p. 179). When compassion fatigue impacts a system, empathy is reduced and some of the distress may be mirrored. All environments can contribute to negative compassion fatigue outcomes. Those environments can be unsupportive and less helpful or actually harmful for the individual experiencing trauma. In this way a diagnosis of PTSD not only implies the individual experiencing symptoms is experiencing some disruption, but may also be harmful to care-takers.

Trauma informed care. In situations when compassion fatigue occurs unchecked it may not be trauma informed. Evans and Coccoma (2014) call for a paradigm shift to recognize the informed understanding of the impact trauma can have on the entire human experience, particularly in the ways trauma is recognized. This calls for the use of best practice methods to respond to trauma whether it is viewed as: a singular isolated event, not an inherent pathological occurrence, or a recognition of all symptoms as a result of trauma exposure. Trauma informed care is compatible with PTSD evaluated within systems theory. The simultaneous interactions occurring and circular causal factors deem the diagnosis as an unnecessary label and narrow way of evaluation. Though the point of origin may be less important and there may be equifinality, responding the best way possible can help improve outcomes. Bloom (2006) does extensive work with trauma informed care and identified that a parallel process occurs amongst those interacting with and experiencing trauma. Parallel process is when symptoms can be passed through the systems in play, similar to Basham's mention of transmission. Even though systems theory can expose some of the complexities of PTSD even that has its limits in providing the clearest picture.

Limitations, Implications and Conclusions

Systems theory has been researched as an efficient and effective intervention for PTSD, but its use as a critique is more nuanced. There is consensus that the diagnosis of PTSD is inadequate or at least incomplete. Zoladz and Diamond (as cited in Evans and Coccoma, 2014) suggest that different sub-types of PTSD have different biological profiles. "The complex interplay between developmental, genetic, endocrine, and neurobiological irregularities found in persons with PTSD indicate that a simplistic diagnostic view of this disorder may need to be re-examined" (Evans & Coccoma, 2014, p. 194). This suggestion is in line with the popular critique of PTSD as reductionist. However, just as the trauma portion can be understood through systems theory, so can resiliency.

Connections to systems can promote resilience as much as those connections can absorb and feel trauma. Human connections are also an important part of self-care and can combat the impact of compassion fatigue. Tosone (2008, p. 75) reports that high numbers of people will and do experience trauma, but the "feelings of commonality and universality" can be restorative to individuals and systems. Placing less focus on differentiating and alienating people based on the severity of his/her traumatic experience may promote resiliency. Harvey (2007) outlines some ecological premises of resiliency to support the shift of focus. Those include understanding resiliency as: transactional and contextual; expressed in varying degrees across different dimensions of functioning; rooted in an enhancement of the relationship between person and context; paying attention to culture; and continued and lasting connections to the environment of the person. Systems theory provides a strong lens to critique the diagnosis of posttraumatic stress disorder, as well as a restorative approach to move in a positive direction.

References

Acevedo, G., & González, M. (2011). Immigration and its effects. In Heller, N., & Gitterman, A. (Eds.), Mental health and social problems: A social work perspective (pp. 425-470). New York, New York: Routledge.

American Psychiatric Association. (2013). Diagnostic and statistical manual of mental disorders (5th Ed.). Washington, DC.

Bernardon, S., & Pernice-Duca, F. (2010). A family systems perspective to recovery from posttraumatic stress in children. The Family Journal, 18(4), 349-357.

Basham, K. (2011). Trauma Theories and Disorders. In Berzoff, J., Flanagan, L. M., & Hertz, P. (Eds.), Inside out and outside in: Psychodynamic clinical theory and psychopathology in contemporary multicultural contexts. (3rd ed., pp. 440-474). Lanham: Rowman & Littlefield Publishers, Inc.

Bloom, S. (2006). Neither liberty nor safety: The impact of fear on individuals, institutions, and societies, part IV. Psychotherapy and Politics International, 4(1), 4-23. Retrieved March 14, 2015, from www.interscience.wiley.com

Bride, B. (2012). Social work contributions to trauma research. In Figley, C. (Ed.), Encyclopedia of trauma an interdisciplinary guide. Thousand Oaks, California: SAGE.

Drozdek, B. (2012). Culture and trauma. In Figley, C. (Ed.), Encyclopedia of trauma an interdisciplinary guide. Thousand Oaks, California: SAGE.

Evans, A., & Coccoma, P. (2014). Trauma-informed care: How neuroscience influences practice. New York, New York: Routledge.

Harvey, M. (2008). Towards an ecological understanding of resilience in trauma survivors. Journal of Aggression, Maltreatment & Trauma, 14(1-2), 9-32.

Kirkpatrick, H., & Heller, G. (2014). Post-traumatic stress disorder: Theory and treatment update. International Journal Psychiatry in Medicine, 47(4), 337-346.

Mather, S. (2012). Cultural aspects of trauma. In Figley, C. (Ed.), Encyclopedia of trauma an interdisciplinary guide. Thousand Oaks, California: SAGE.

Miehls, D. (2011). Racism and its effects. In Heller, N., & Gitterman, A. (Eds.), Mental health and social problems: A social work perspective (pp.180-242). New York, New York: Routledge.

Robbins, S., Chatterjee, P., & Canda, E. (2012). Contemporary human behavior theory: A critical perspective for social work. (3rd ed., pp. 25-58). Boston: Allyn & Bacon.

Rothery, M. (2008). Critical ecological systems theory. In N. Coady & P. Lehman (Eds.), Theoretical perspectives for direct social work practice: A generalist-eclectic approach (pp.89-118). Springer Series on Social Work. New York, NY: Springer Publishing Company LLC.

Tosone, C. (2008). Shared trauma. Psychoanalytic Social Work, 10(1), 57-77. Retrieved March 14, 2015, from http://www.tandfonline.com/loi/wpsw20

Van der Kolk, B. (2002). Posttraumatic therapy in the age of neuroscience. Psychoanalytic Dialogues, 12(3), 381-392

A Feminist Critique of Diagnosis in the DSM 5: *Queering* Gender Dysphoria

Jessica Piña

Introduction

Feminist scholars and queer theorists, among other promoters of social justice and advocates of marginalized populations, have long challenged psychiatric diagnoses, psychological theories, and clinical practices, particularly those in the *Diagnostic and Statistical Manual of Mental Disorders* (DSM; Marecek & Gavey, 2013, p.3). In this paper I will use a critical lens informed by feminist and queer theory to explore the ways in which Gender Dysphoria, as described in the DSM fifth-edition (DSM 5; American Psychological Association, 2013), is not only ambiguous and contradictory, but can also be oppressive and harmful to those associated with the diagnosis. My primary concern is to expose the ways gender dysphoria in the DSM preserves a patriarchal social order and further perpetuates power imbalances along gendered lines. Consequently, I pose the following questions: What does Gender Dysphoria in the DSM inflict on knowers and on clinical knowledge about this population? What prevalent moralities and norms regarding gender and sexual identity does the diagnosis of Gender Dysphoria preserve? Lastly, I propose that a feminist and queer approach to research methods and practice can serve as effective tools to deconstruct the binary and static knowledge produced by categories such as Gender Dysphoria.

The DSM uses a categorical approach to the diagnosis of mental disorders, it sets criteria that individuals must meet to in order for their symptoms to be legally recognized as a diagnosis. However, the DSM does much more than determine who is mentally ill and who is not. "It serves as the primary determinant of insurance eligibility, disability payments, and who receives special educational services. It figures in legal decisions regarding child custody disputes and criminal responsibility. It shapes the direction of research, the allocation of research funds, and the approval of new drugs" (Merecek & Gavey, 2013, 4). More importantly, I argue the DSM serves as a manual for maintaining a Western, heteronormative status quo, and is therefore problematic. To start, the simple act of labeling an individual as "abnormal" shifts the power dynamic from the individual to the mental health professional in determining "normalcy" (Anderson, Cermele, & Daniels, 2001, p. 230). In this way, those who deviate from the 'norm' become more vulnerable to marginalization and societal alienation. In this next section, I will review relevant feminist and queer theory principles in order to critically examine Gender Dysphoria in the DSM 5.

Feminism And Queer Theory

Queer Theory and its Beginnings

Prior to and up until the early twentieth century, the term *queer* was derogatory and offensive as well as an umbrella term for those identifying with the LGBTQIA community (Carlin & DiGrazia, 2004, p. 1). Today the term queer has not only evolved in definition but has also become a place of transforming and of constant deconstructing of norms. Many scholars and theorists from various disciplines advocate for what is known as "Queer Theory." For these scholars 'queer' is not a term that should function as an umbrella for LGBTQIA studies or activism (Fotopoulou, 2012, p. 25). Queer theory emerged in the 1990's out of two stands; one from feminist scholar Judith Butler, who argues that sexuality is culturally produced and further deconstructed the link between gender and sexuality, (Fotopoulou, 2012, p. 25) and the other from Lesbian and Gay Studies. Furthermore, queer theory denies the existence of any fixed, universal, absolute truth across cultures, and thus, claims that identities are fluid and ever changing (Carlin & DiGrazia, 2004, p. 1). For the purpose of this paper, I will use Eve Sedwick's proposal for the use of queer theory. Eve Sedwick, one of queer theory's most important advocates, argues that our gender choice has been the essential factor in the organizing of social relations in Western civilizations. Sedwick proposes that we deconstruct gender in the categorization of our social relations through an analysis of the historical process by which binary constructions began, secondly, through an analysis of how this binary value system operates and lastly, by exploring ways to challenge and disrupt the binary itself (Carlin & DiGrazia, 2004, p. 2).

Feminist and queer methodological frameworks do not come with a set of tools or processes to do research. They serve as critical and intersectional lenses in which to analyze our current systems of power. By applying queer theory's claim that identity, specifically gender and sexual identity, is permeable and provisional, gender dysphoria as a diagnosis can be disregarded as unjust to those who identify with any sexual identity not described as *normal*. "Queer theory, when applied as a distinct methodological approach to the study of gender and sexuality, has sought to denaturalize categories of analysis and make normativity visible" (Fotopoulou, 2012, p. 19). *Queering* gender dysphoria requires that we not only examine the diagnostic criteria posed by the DSM, but also deconstruct the methods of research that led to such criterion.

Feminist Foundations

As a feminist, it is important to identify power relationships and critically account for the ways in which they empower the privileged and oppress vulnerable populations (i.e. people of color, people with disabilities, women etc.). Further, feminists assert that critical lenses that allow us to challenge privileged narratives and deconstruct the normativity of society are crucial to achieving social change (Fotopoulou, 2012, p. 22). "The cultural, local and historical production of knowledge, is key to the production of self-critical and accountable feminist theory; *how* we do research generates relationships through and in the selective steps that determine which differences matter" (Fotopoulou, 2012, p. 24). Additionally, feminist perspectives in social research question positivism's answers to the epistemological questions of who can possess knowledge, how knowledge is obtained, and what knowledge is (Hesse-Biber, Leavy, Yaiser, 2004, p.11). Many feminists conceptualize *truth* differently than mainstream researchers and assert that marginalized groups can possess knowledge. They also recognize that research methods vary, and no one method is inherently better than another although many critique positivism and traditional research methods as intrinsically patriarchal (Hesse-Biber, Leavy, Yaiser, 2004, p.11).

As feminists, we must critically examine what historically and currently is accepted knowledge. Doing so exposes its limitations and is important "because different methodologies produce different knowledges and hence can offer different political visions" (Fotopoulou 21). This allows us to question gender dysphoria presented as an absolute truth about the transgender population and examine the factors and values that assisted in its development.

A Feminist Look At Gender Dysphoria

"In 1980, Gender Identity Disorder in Children (GIDC) was introduced in the third edition of the DSM (American Psychological Association, 1980)… for adults, the DSM-III introduced the category of Transsexualism under the broader category of Gender Identity Disorders (GID). This classification represented the premiere of transsexualism as an official disorder" (Sennott, 2011, p. 95). In 1994, the DSM-IV was published (American Psychological Association), wherein Gender Identity Disorder became diagnosable in both children and adults. Today, the DSM 5 uses the term "Gender Dysphoria" as a more descriptive version of the previous disorder (GID, DSM 5, American Psychological Association, 2013, p. 415).

Revealing the criteria for Gender Dysphoria is important because much of the controversy that "surrounds the pathologization of gender nonconformity concerns language" (Sennott, 2011, p. 95). More important than relaying information, is raising awareness of the factors contributing to such diagnoses and legislative policies related to gender identification, treatment and rights. Such political outlets have allowed for a more "comprehensive lens through which to critique the diagnosing of gender differences" (Sennott, 2011, 97).

The DSM claims that "Gender Dysphoria" as a diagnosis, refers to the *"distress* that may accompany the incongruence between one's experienced or expressed gender and one's assigned gender" (DSM 5; American Psychological Association, 2013, p. 452). Unlike the DSM-IV, the DSM 5 states gender dysphoria now "focuses on dysphoria as the clinical problem, not identity per se" (DSM 5; American Psychological Association, 2013, p. 451). However, the diagnostic criteria accompanying the illness generally consist of the various ways an individual can display feelings related to the desire to be of a different gender (American Psychological Association, 2013, p. 451). Subsequently, the identity behaviors themselves become an index of Gender Dysphoria disorder. It is important to note basic definitions currently accepted in order to better understand what the diagnosis is referring to. To begin, the word *dysphoria* is defined as a profound state of dissatisfaction with self (Stevenson, 2010). Secondly, transgender, as defined in the DSM 5, is relating to a person whose identity does not correspond to that person's birth sex.

Disturbingly, the diagnostic criteria of gender dysphoria, aligns the *dysphoria* and *distress* only as accompanying one's incongruent identity experience (DSM 5; American Psychological Association, 2013, p. 452). Categorizing the distress that may come from a transgender experience as a mental disorder insinuates that there should be an otherwise 'normal or acceptable' response, if one at all. In addition, by using a categorical approach to diagnosis based on the presence or absence of specified symptoms, "the DSM essentially takes symptoms out of context for examination and categorization, and thus become an illustration of the way gender and race are constructed in theory and practice" (Anderson, Cermele, & Daniels, 2001, p. 233).

As described in the DSM, the distinction between dysphoria/distress and the experience of a transgender person is blurred, ambiguous and left for misinterpretation. As a result, the categorical criteria create an *essentialized* description of a one's experience being transgender in addition to feeling distress. Labeling a person's experienced gender identity (whether congruent to birth sex or not) as a mental illness is disempowering and stigmatizing.

Furthermore, the historical inclusion of gender dysphoria as a diagnostic category indicates normative definitions of health are limited to heteronormative identities and reflects the construction of the experience of being transgender and feeling distressed as a mental illness and in need of medical or psychiatric intervention (Anderson, Cermele, & Daniels, 2001, p. 232). The pathologizing of people identifying as transgender suggests those that identify as cisgender are the prototype of normality. Gender dysphoria thus "hinders the autonomy of gender non-conforming persons who do not feel impairment or distress due to their gender identification. Often, these unaffected identities which are emerging in the 21st century do not wish to 'transition' from one end of the gender," (Sennott, 2011, p. 101). A more inclusive goal, seen through a queer lens, would be to find a position on the gender identity spectrum where people feel the most psychologically at ease (Sennott, 2011, p. 102).

Queering Gender in Gender Dysphoria

As previously mentioned, *queer* stands against homogenizing and contests normativity. This term allows us to *become* queer, a way of deconstructing assumed categories of analysis. Consequently, the existence of gender dysphoria in the DSM represents a highly controversial issue in today's society and is problematic for many reasons. For one, the terminology used in the gender dysphoria diagnosis, reifies the binary gendered system we live in and value, i.e. male/female, man/woman, and masculine/feminine. For example, in the "Diagnostic Features" under Gender Dysphoria, girls who display gender dysphoria are described as wanting to play contact sports and refusing to play with dolls (among other things) (DSM 5; American Psychiatric Association, 2013, p. 453), the underlying assumption is that a girl who likes contact sports is not 'normal.' In this sense, the privileged identities are those that are "invisible" and thus unspoken of, i.e. a person who is designated female at birth, becomes a woman, expresses her gender with feminine characteristics, and is attracted to a designated male at birth, man, who displays a masculine gender expression etc.

The feminist approach operates with the understanding that we construct our own gender identities as relating to our surroundings. Essentializing gender identity can be just as dangerous as resorting to biological essentialism, i.e. there is not one truth about our sexual identity or biology. In order to better understand the ways in which feminist perspective counters gender dysphoria I will use Shannon Sennott's (2011) foundational principles of her proposed transfeminist approach: "1. There does not exist a hierarchy of authentic lived experience for women; 2) To privilege one type of womanhood or femaleness over another is inherently anti-feminist; 3) No one individual, group, or type of woman is permitted to define what it means to be a woman; and 4) Most trans and/or gender nonconforming individuals have had lived experience, either past or present, as a girl or woman and have suffered the direct repercussions of socially condoned misogyny and systemic gender based oppression" (p. 102).

Congruent to a gendered self is a need for gender expression. It is to be expected that to be denied gender expression, or any other aspect of our lives may typically lead to some form of psychological response, i.e. stress, anger, etc. By suggesting that this response is a mental disorder, the DSM reinforces negative stereotypes of people with variant gender identities. Moreover, "as agents of social control, patriarchal psychomedical institutions have diagnosed gender differences to ensure sociopolitical homeostasis and maintain disciplinary authority" (Sennott, 2011, p. 95).

Another aspect to consider in Gender Dysphoria as a diagnosis is the disregard for the importance of multiple factors (i.e. cultural, environmental, and socio-economic etc.) impacting a person's identity. "Feminist psychologists have always insisted that psychological suffering is linked to the broad social, economic and political context. Distress and emotional wellbeing, as well as disadvantage and privilege, are distributed along intersecting axes of gender, racial/ethnic categorization, economic status, and sexuality" (Marecek & Gavey, 2013, p. 7).

Traditionally, clinicians complain that clinical research does not speak to the complexity of patients' environments and are therefore slowly responsive to research findings (Swartz, 2013, p. 47). Understanding the impact of context, social–cultural, interpersonal relationships, etc. is fundamental in examining a person's mental health state. Behaviors and responses to an experienced identity- "extracted from social context, can be viewed as a straightforward outcome of biological forces, which sometimes go awry in their object choice" (Duschinsky & Chachamu, 2013, p. 53). Furthermore, the fact that disorders are given a "culture-related diagnostic issues" section is reflective of the assumption and value that Western/white culture is universal.

Conclusion

Queering social research and deconstructing the knowledge it generates requires attention to research methodologies and processes. Further research would require creating new methods of collecting and conducting research in a way that does not maintain the status quo. For example, exclusions of certain groups may manifest how groups are privileged over others. "Browne (2007) argued that even though sexualities are included as categories of difference in research, these categories are normalizing. These common attributes cannot be objectively observed and statistically measured" (Fotopoulou, 2012, p. 26). In other words, there is a 'normal' created during the research process that is then reaffirmed with the use of statistical data. Anderson and Cermele analyzed case studies in the DSM IV Casebook to identify differences in the quantity and quality of physical and sexual details, they argued the case studies contribute to a gendered and raced understanding of mental illness and mental health (Anderson & Cermele, 2001, p. 230). This is significant to consider as it suggests the importance of examining the ways the DSM Casebook constructs gender and race in the context of mental illness, and we can further assess the impact these assumptions have in the Diagnostic and Statistical Manual of Mental Disorders.

As a feminist and social justice advocate, I believe further research and study should involve the true immersion in the LGBTQIA community in order to better assess needs and decisions regarding the future of Gender Dysphoria in the DSM. We must always critically think of the ways mental diagnosis and illnesses are *othering* and privileging. Maintaining Gender Dysphoria in the DSM 5 perpetuates binary norms of sexual identity and expression and cannot exist without the consideration of all the intersecting factors on the identity.

References

American Psychiatric Association. (2013). Diagnostic and Statistical Manual of Mental Disorders (5th ed.). Washington, DC.

Carlin, D., and DiGrazia, J. "What Is Queer Theory." Queer Cultures. Upper Saddle River, NJ: Pearson/Prentice Hall, 2004. Pgs.1-2. Print

Cermele, J., Daniels, S., & Anderson, K. (2001). Defining Normal: Constructions of Race and Gender in the DSM-IV Casebook. Feminism & Psychology, 11(2), 229-247.

Duschinsky, R., & Chachamu, N. (2013). Sexual Dysfunction and Paraphilias in the DSM 5: Pathology, Geterogeneity, and Gender. Feminism & Psychology, 23(1), 49-55.

Fotopoulou, A., (2012). Intersectionality Queer Studies and Hybridity: Methodological Frameworks for Social Research Journal of International Women's Studies, 13(2), 19-31.

Hesse-Biber, S., & Yaiser, M. (2004). xxxx. In Hesse-Biber, S. & Yaiser, M. Feminist Perspectives on Social Research (p. 101-120). New York: Oxford University Press, Inc.

Marecek, J., & Gavey, N. (2013). DSM 5 and beyond: A critical feminist engagement with psychodiagnosis. Feminism & Psychology, 23(1), 3-9.

Sennott, L. S., (2010) Gender Disorder as Gender Oppression: A Transfeminist Approach to Rethinking the Pathologization of Gender Non-Conformity. Women & Therapy, 34(1-2), 93-113.

Stevenson, A. (2010). Oxford dictionary of English (3rd ed.). New York, NY: Oxford University Press.

Swartz, S. (2013). Feminism and psychiatric diagnosis: Reflections of a feminist practitioner. Feminism & Psychology, 23(1), 41-48.

Culture and the DSM

Hina Rehman

Introduction

The lack of culture competency is one of the biggest issues of the United States, although the United States is considered to be one of the most diverse and well established countries. The deficiency of understanding different cultures seems to be an ongoing problem. The second largest city, Chicago, is one of the most segregated cities in terms of class and race. The most standard segregation measure shows that American cities are now more segregated than they have been since 1910 (Glaeser&Vigdor 2012). The unfortunate part is that with all the class and racial segregation it seems almost impossible to learn about different cultures. The incline in segregation can be partly attributed to the reform of government verdicts (Glaeser&Vigdor 2012). The educational systems seem to only focus on the history of the United States and how rich white men saved the day rather than learning about the different cultures (Fine, 1990).

In this paper, I will address issues that have come across in the *Diagnostic Statistical Manual of Mental Disorders* (DSM). First I will highlight two theories that are designed to see through the perspective of an environment (PIE) and social cultural lens. Second, I will explain what the DSM consists of and how clinicians determine their diagnosis. Finally, I will discuss the factors of three specific Trauma-and-Stressor-Related Disorders and how they can be misdiagnosed through uneducated views on culture.

Person in Environment Theory

PIE's roots start with early debates that clinicians focus their attention only on the individual rather than the environmental impacts on the individual (Krondat, 2002). A PIE perspective provides adequate framework for assessing an individual in addition to his or her present problem and assets rather than focusing solely on changing a person's psyche (Kondrat, 2002). The perspective of PIE concentrates on the notion that an individual and his or her behavior cannot be understood sufficiently without consideration of the various aspects from that individual's environment. This may include but not limited to the following aspects: social, political, familial, temporal, spiritual, economic, cultural, and physical (Krondat, 2002).

In many cultures depression does not exist. Different cultures and cultural norms judge issues differently. Some cultures teach people to repress their memories and live with unsolved problems that can lead into mental illnesses. In many minority groups, children of all classes tend to be hypersensitive and anxious about their relations with different societies (Clark,2004). Similarly, in my clinical profession I have seen African American women compare their problems to their ancestors with the idea of "If they can get through slavery, then what we're going though is nothing." This idea results from the lack of education students receive about different races. There are many theories that study the different facets of an individual's life. For the purpose of this paper I will focus on PIE and Social Constructionism.

Social Constructionism

Social Constructionism focuses on how a sociocultural and historical context shapes individuals as well as creation of knowledge and how individuals create themselves (Hammel,2009). Social constructionism also upholds the explanation of how culture can have an impact on how individuals interpret events. It is important to understand how individual experiences are subjective to their statistic process. Humans are self-interpreting beings (Hammel,2009). Each individual's life experience and historical context can impact their view on mental health or any physical diagnosis.

The ideal of cultural competence was initially perceived as knowledge of the "values, customs, and traditions" of a specific cultural group (Hammel,2009). Unfortunately, the ideal culture in the United States is viewed on what historical Caucasian males have deemed to be normal. This ideal was a result of the assumption that culture was experienced and interpreted in the same way within a certain group regardless of status, age, gender, race or sexuality. The inability to purchase basic needs is a hazardous aspect of life in poverty and is the source of psychological and physiological suffering for minorities (Smith,2013). This explains that cultures are static and fixed and that the "nuanced" values and perspectives of a culture might be learned by a foreigner (Hammel,2009).

Diagnostic Statistical Manual-5

In the concept of psychiatry and clinical psychology the dominant framework for theorizing psychopathology is coded by the _Diagnostic and Statistical Manual of Mental Disorders_ (DSM) of the American Psychiatric Association (Wayland &O'Brein, 2013). This DSM Manual for most clinicians helps guide defining a mental health diagnosis, for example Trauma Stress Related Disorder, Dissociative Disorder and Neurocognitive Disorder. The diagnosis has several distinguishable features of individuals that help outline the specific diagnosis. Some of these features include, hallucinations, depressed mood and social neglect. Together, this guide helps clinicians label a client's mental problems. According to Kendler et al. the DSM provides clinicians with diagnostic criteria that attains acceptable levels of reliability (Kendler, Kupfer , Narrow, Phillips, & Fawcett 2009.). Furthermore, the current DSM criteria have been criticized by clinicians for not capturing the clinical complexities of many of their clients, such as epidemiological, neurobiological, cross-cultural, and basic behavioral research (Kendler et al, 2009).

There are many clinicians who use the DSM to reassure clients that what they are experiencing is something "real." In other cases the DSM diagnosis helps professionals not with only human behavioral therapy but also with pharmaceuticals. It is important to understand that diagnosing a client not only eliminates their understanding of the other aspects in their life but can be dehumanizing for some clients (Wayland &O'Brian,2013). Furthermore, I will address how the DSM disregards environmental and cultural factors that compute in a mental health verdict.

Criteria of Trauma and Stressor- Related Disorders

Post-Traumatic Stress Disorder

Criterion A: stressor

> The person was exposed to: death, threatened death, actual or threatened serious injury, or actual or threatened sexual violence, as follows: **(one required)**

> Direct exposure.

> Witnessing, in person.

Indirectly, by learning that a close relative or close friend was exposed to trauma. If the event involved actual or threatened death, it must have been violent or accidental.

Repeated or extreme indirect exposure to aversive details of the event(s), usually in the course of professional duties (e.g., first responders, collecting body parts; professionals repeatedly exposed to details of child abuse). This does not include indirect nonprofessional exposure through electronic media, television, movies, or pictures.

Criterion C: avoidance

Persistent effortful avoidance of distressing trauma-related stimuli after the event: (one required)

Trauma-related thoughts or feelings.

Trauma-related external reminders (e.g., people, places, conversations, activities, objects, or situations).

Reactive Attachment Disorder

Criterion B: A persistent social or emotional disturbance characterized by at least two of the following:

Minimal social and emotional responsiveness to others

Limited positive affect

Episodes of unexplained irritability, sadness, or fearfulness that are evident even during nonthreatening interactions with adult caregivers.

Disinhibited Social Engagement Disorder

Criterion A: A pattern of behavior in which a child actively approaches and interacts with unfamiliar adults and exhibits at least 2 of the following:

Reduced or absent reticence in approaching and interacting with unfamiliar adults.

Overly familiar verbal or physical behavior.

Diminished or absent checking back with adult caregiver after venturing away, even in unfamiliar settings.

Willingness to go off with an unfamiliar adult with minimal or no hesitation.

In this section I will review some of the controversies and the assessment instruments currently used to decide whether an individual has experienced a traumatic event after reading the symptoms that form into a diagnosis of an individual client. It appears that most of the symptoms are just synonyms of each other and can be inferred as misleading. These symptom clusters seem effective at explaining the central difficulties of those exposed to singularly occurring, acute traumatic events. However, in seclusion they are less well-suited for the spectrum of symptoms and personality disruption often exhibited by individuals who have experienced prolonged trauma (Dyer, Dorahy, Hamilton, Corry, Shannon, MacSherry, &McElhill, 2009).

Also note that each of the symptoms may have interpretable terminology. An example of this would be, "problems with concentration." It is important to consider that each client may have a different understanding of what they define as concentration; what one may infer as multitasking to another it may be determined as not concentrating at all. To differentiate how each client may interpret each symptom can be a factor in their environment and/or culture. The environment follows the aggregate of surrounding things, conditions, or influences, surroundings; milieu. (Environment, 2014).

Above displayed is some criterion factors of PTSD. PTSD not only disregards the culture aspect of trauma related features but needs to understand better causes and insinuations of the differences found in the conditional panorama of PTSD across cultures (Friedman,2014). Stress traumatic disorders are usually misinterpreted because clinicians may not understand the value of certain cultures.

Criteria A of PTSD states "Exposure to actual or threatened death, serious injury, or sexual violence..." (American Psychiatric Association, 2013). By what means is an individual supposed to understand what a threatened death or sexual violent behavior consists of? In some cultures threats are made every day, and it is difficult for the client to distinguish the difference between an actual death-threat and a non-death threat. With this ambiguity, it is hard to determine what is considered a traumatic event.

A solid example of this reasoning is within gang culture. When joining a gang, often times there is an initiation that needs to be passed. This initiation is usually a violent crime that could include "theft, murder, gang-rape, or drive-by shootings." Therefore when questioning individuals in gangs the term threatened death or death may not be considered a traumatic event. Another example is losing a child. Women suffering miscarriages due to drug and alcohol use, heavy lifting, and or too much physical work could be considered witnessing a death. However, the DSM does not define whether a miscarriage or abortion is considered to be a "death."

Additionally, views on sexual violence can vary amongst different cultures. In many societies, women and men regard marriage as entailing the obligation on women to be sexually available virtually without any limit. Even outside of a marriage, some women may have few reasonable options to refuse sexual advances (WHO,2013). When discussing sexual assault as a learned behavior, it is important to consider social conditions, norms, rules and prevailing attitudes about sex which frame the structure of rapist within the broader perspective of an individual's environmental system (WHO, 2013). In many societies women are bound to sexually please their man with or without their consent. Again, distinguishing what may be a traumatic event could be reasonably unreliable through the DSM.

Criteria A.3 for PTSD says "Indirectly, by learning that a close relative or close friend was exposed to trauma. If the event involved actual or threatened death, it must have been violent or accidental" (American Psychiatric Association, 2013). In what case does the client or clinician determine what an accidental and violent death is? Are people who get diagnosed with cancer and decease supposed to be sectioned into a violent or accidental death? An individual living within a poverty line household, with no health insurance, dying of cancer can be considered traumatic to many people. Thus, interpreting what an accidental death to a client may be unreasonable.

Criterion C.2 of PTSD displays "Persistent effortful avoidance of distressing trauma-related stimuli after the event: trauma-related thoughts or feelings"(American Psychiatric Association, 2013) . Thus an individual avoiding any negative thoughts that may or may not be related to the traumatic event can easily be interpreted to fall under the range of displaying a feature of PTSD. In my professional clinical experience I have worked with African American clients in impoverished neighborhoods. Most of the clients have witnessed a death or hear about a death as often as every week. Therefore when attending events outside of their neighborhood (school work, etc.) most clients like to avoid thinking about where they live. Not only because of the deaths occurring in the neighborhood or the violence that occurs, but because of the scenery it displays. Homes with broken windows or not having a doctor in reach can be very frustrating for individuals and may be avoiding negative thoughts.

Criterion B. 1 states "Minimal social and emotional responsiveness to others" (APA), as mentioned earlier although the United States is considered to be one of the most diverse countries globally some of the biggest cities are the most segregated (Census). When tying in person in environment theory and social constructionism it becomes easy to understand that an individual's apprehend on cultures is easily affected by where they grow up. Therefore having "minimal social and emotional responsiveness to others [in different cultures]" can be viewable as a fault in not having the knowledge about different cultures rather than receiving a diagnosis. This can simply be because of the lack education taught in schools about culture.

Criterion A.1 of Social Engagement Disorder declares "Reduced or absent reticence in approaching and interacting with unfamiliar adults". In the US it is normal for men to shake hands when they meet other men but is very unusual for men kiss one another as a greeting. However, in France, individuals including school children, shake hands with their friends or kiss them on both cheeks, each time they meet and leave. Also in Polynesia you take your friends hands and use them to stroke your face (Cairns Regional council). In American culture things of this nature would be considered very unusual and actually a symptom of a disorder. Especially with children who are not able to explain themselves. Moreover, criterion A.3 of Social Engagement Disorder reads "Diminished or absent checking back with adult caregiver after venturing away, even in unfamiliar settings" (Smith, 2013). In some countries, checking back with an adult in any setting can be considered a sign of respect.

Conclusion

After presenting PIE and Social Constructionism in relation to the DSM is it easy to understand that the DSM does not base its diagnosis on the understanding of different cultures. With using only three disorders I analyzed how culture is not taken in context when it comes to the definitions of diagnosis. This affects everyone as a whole in society and should be blamed on the American education system. The educational systems need to be changed. This should happen by taking different cultures into context in syllabus of teachers.

The United States is filled with different cultures and races. This should be taught in our educational systems regardless of socio economic status. This will help improve the understanding of different cultures and lessen several diagnosis.

References

American Psychiatric Association. (2013) *Diagnostic and statistical manual of mental disorders*, (5th ed.). Washington, DC: Author.

Clark, K., Chein, I., & Cook, S. (2004). The Effects Of Segregation And The Consequences Of Desegregation A (September 1952) Social Science Statement In The Brown V. Board Of Education Of Topeka Supreme Court Case. *American Psychologist*, 495-501.

Cultural Greetings. (n.d.). Retrieved April 17, 2015, from http://www.cairns.qld.gov.au/__data/assets/pdf_file/0007/8953/CulturalGreetingExercise.pdf

Dyer, K. W., Dorahy, M. J., Hamilton, G., Corry, M., Shannon, M., MacSherry, A., &McElhill, B. (2009). Anger, aggression, and self-harm in PTSD and complex PTSD. Journal Of Clinical Psychology, 65(10), 1099-1114

Environment. (2014.) *American Heritage® Dictionary of the English Language, Fifth Edition*. (2011). Retrieved from http://www.thefreedictionary.com/environment

Fine, M. (n.d.). "The Public" in Public Schools: The Social Construction/Constriction of Moral Communities. *Journal of Social Issues*, 107-119.

Friedman, M. (2014). *Handbook of PTSD: Science and practice*. New York: Guilford Press.

Glaeser, E., &Vigdor, J. (2012). The end of the segregated century: Racial Separation in America's Neighborhoods, 1890-2010. *Manhattan Institute For Policy Research, 66*.

Hammell, K. (2009.). Occupation, well-being, and culture: Theory and cultural humility / Occupation, bien-etre et culture : La theorie et l'humiliteculturelle. *Canadian Journal of Occupational Therapy*, 224-234.

Kendler, K., Kupfer, D., Narrow, W., Phillips, K., & Fawcett, J. (2009.). Guidelines for making changes to the DSM 5. 1-10.

Kondrat, M. E. (October 01, 2002). Actor-Centered Social Work: Re-visioning 'Person-in-Environment' through a Critical Theory Lens. *Social Work, 47*, 4.)

Smith, L. (2013). So close and yet so far away: Social class, social exclusion, and mental health practice. *American Journal of Orthopsychiatry*, 11-16

Wayland, K., &O'Brein, S. (2013). DSM 5 Changes to the DSM and how they will affect our clients with mental impairment. *Deconstructing Prejudicial Psychiatric Labels*, 1-30.

WHO. (2013). In *World Report on violence and health* (Etienne G. Krug, Linda Dahlberg, James Mercy ed., p. 72). WHO.